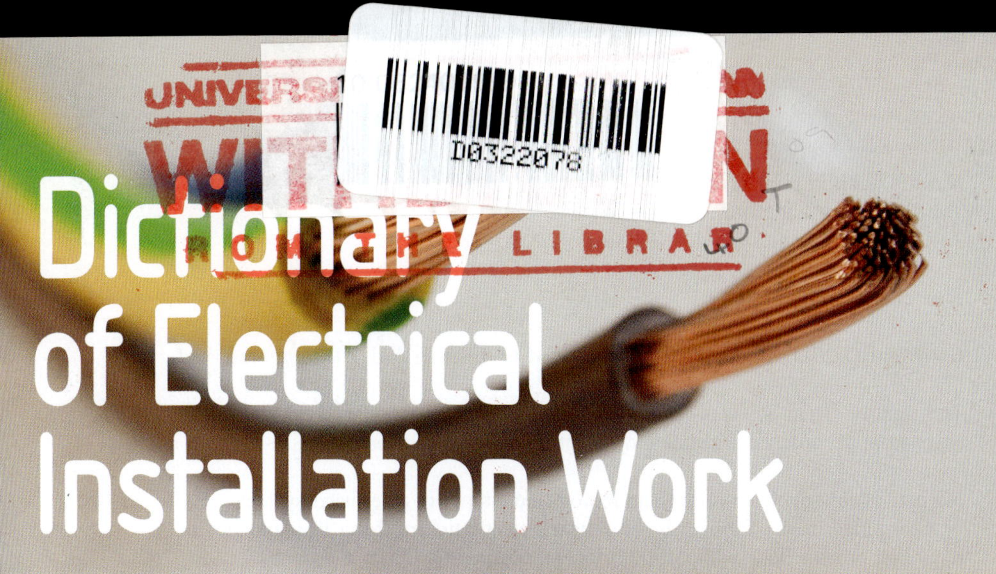

Dictionary of Electrical Installation Work

Illustrated Dictionary– A practical A–Z guide

Brian Scaddan

Amsterdam • Boston • Heidelberg • London • New York • Oxford
Paris • San Diego • San Francisco • Singapore • Sydney • Tokyo
Newnes is an imprint of Elsevier

Newnes is an imprint of Elsevier
The Boulevard, Langford Lane, Kidlington, Oxford OX5 1GB, UK
30 Corporate Drive, Suite 400, Burlington, MA 01803, USA

First edition 2011

British Library Cataloguing in Publication Data
A catalogue record for this book is available from the British Library

Library of Congress Cataloging-in-Publication Data
A catalog record for this book is availabe from the Library of Congress

ISBN: 978-0-08-096937-4

Printed and bound in Italy

11 12 13 14 15 10 9 8 7 6 5 4 3 2 1

Introduction

Over the years I have encountered many occasions when electrical operatives use words or phrases that are either incorrect or are not fully understood. In this dictionary I have included entries that relate to electrical installation work, both theory and practice. There is also a section devoted entirely to formulae.

This book should provide a useful accompaniment to other text books and guides, and will also act as a valuable 'stand alone' reference source for both qualified electrical personnel and students alike.

Brian Scaddan

A

a.c. (alternating current)

This is usually produced by a.c. generators, but it may be derived electronically from a direct current (d.c.) source such as a photo-voltaic (PV) solar panel by use of a **d.c.** to **a.c.** PV invertor.

Fig. 1 shows the sine wave for a typical 230 V a.c. supply from the Distribution Network Operator (DNO).

The frequency of the supply is 50 cycles-per-second or Hertz (50 Hz).

The UK supply voltage is 230 V, +10%/–6% giving a range of 216.2 V to 253 V.

The r.m.s. (root mean square) value of current (ampere) gives the same heating effect as a similar value of d.c. current, so 10 A a.c. r.m.s. will cause as much heat as 10 A d.c.

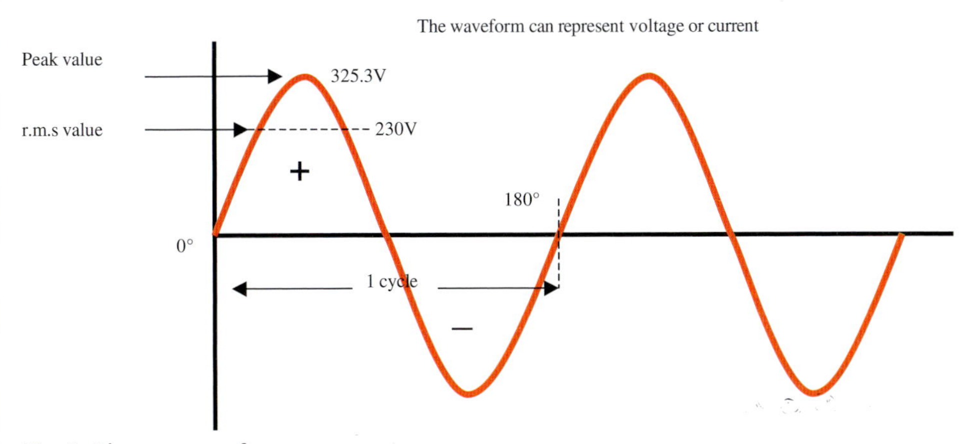

Fig. 1 The a.c. waveform

The Dictionary of Electrical Installation Work. DOI: 10.1016/B978-0-08-096937-4.00001-5

The rms value of an alternating quantity occurs at the point when a generator has moved through a rotation of 45°.

Unless otherwise stated, values quoted are rms values.

Accessory

(From BS 7671:2008 definition) 'A device, other than current-using equipment, which is associated with such equipment or with the wiring of the installation.' So, anything such as socket outlets, lampholders, distribution boards, emergency stop buttons, etc, etc, etc.

Additional protection

This is extra protection against electric shock and is provided by:

1. RCDs with a rating $I_{\Delta}n$ not exceeding 30 mA and an operating time of 40 ms at a residual operating current of $5\,I_{\Delta}n$, or *(see also Residual current devices)*
2. Supplementary equipotential bonding *(see under Earthing).*

Additions and alterations

An addition extends or adds to an installation, e.g. a spur from a ring circuit; an extra lighting point; a new motor circuit, etc.

An alteration is a change to an existing installation, arrangements, e.g. new for old; consumer unit change; re-positioning an accessory, provided the cable length is not increased as this would technically be an addition.

No addition or alteration should impair the safety of the existing installation or, conversely, have its safety impaired by the existing.

For example:

1. A new spur from a socket outlet circuit may not be safe if the loop impedance (Zs) of the existing circuit is near the permitted maximum.
2. A class 1 light fitting should not replace an old fitting that is supplied by a cable with no circuit protective conductor, unless the replacement was made because the old fitting was damaged. In this case it could be argued that the replacement would leave the situation in a safer condition.
3. An extra load, e.g. a new 10.5 kW shower circuit, could result in the maximum demand being exceeded, causing overloading of main tails, metering, etc.
4. Changing a consumer unit housing BS 3036 fuses to one with BS EN 60898 circuit breakers would require a thorough test and inspection of the existing installation, to ensure it was safe for the change and that it did not impair the existing. For instance a 5 A BS 3036 fuse protecting a lighting final circuit has a tabulated maximum Zs value of **9.58 Ω** and the nearest equivalent is a 6A BS EN60898 type B which has a tabulated maximum value of **7.67 Ω**.
   This means that if the circuit had an actual value of say 8 Ω, the BS 3036 fuse would operate, in the event of a fault, within the maximum permitted time but the change to the circuit breaker would result in a shock risk condition. This situation could be overcome by using an RCBO.

Adiabatic equation

The word adiabatic means the reduction or absence of heat transfer. The equation enables a suitable conductor size to be chosen to ensure that it will not be damaged by heat due to excessive fault current. The equation is:

$$S = \frac{\sqrt{I^2 t}}{k}$$

Where S = cross section area of the conductor (mm²)

I = fault current (A)

t = duration of fault

k = factor taken from tables and depends on conductor and insulation material.

(see also Circuit protective conductor and Let-through energy)

ADS (see Automatic disconnection of supply)

Agricultural/horticultural locations

(BS 7671:2008 Section 705) These include livestock and arable farms, but not farm dwellings, stables, garden centre glasshouses, greenhouses, etc.

Main Points:

- Fuses and circuit breakers are used for overload and short circuit current protection
- RCDs for earth fault protection (regardless of the earthing system):
  1. 30 mA or less for socket outlet circuits up to 32 A
  2. 100 mA or less for socket outlet circuits over 32 A
  3. 300 mA or less for all other circuits and for fire protection
- Equipment to be rated at a minimum of IP 44
- Where agricultural vehicles and machinery are used:
  - Underground cables need to be at a depth of at least 600 mm with extra mechanical protection, and where crops etc. are grown, at least 1.0 m deep
  - Self-supporting suspension cables should be at least 6 m above ground level
  - Conduit and trunking systems should be able to resist an impact of 5 joules *(see IK codes)*
- Supplementary equipotential bonding must be provided between all exposed and extraneous conductive parts accessible to livestock.

Alterations (see Additions and alterations)

Ambient temperature

(BS 7671:2008 definition) 'The temperature of the air or other medium where equipment is to be used.'

The standard air temperature for cable current carrying capacity is 30°C.

The standard ground temperature for underground cable current carrying capacity is 20°C.

At these temperatures no adjustment to tabulated cable current rating is necessary.

Ampere symbol A

This is the unit of electrical current, which is named after the French physicist André Marie Ampère (1775–1836).

Amusement parks (see Fairgrounds)

Architectural symbols (see Diagrams)

Arm's reach

This is reaching with either arm, without assistance, to a distance of 2.5 m from a standing position and 1.25 m downwards from a lying position.

Placing out of arm's reach is permitted as a method of preventing contact with live parts (basic protection) but only when the installation is under the control or supervision of skilled persons.

An example of this, for instance, would be that of an overhead travelling crane in a factory where it derives its electrical motive power from the rails it runs on. Clearly these live rails must not be within arm's reach.

ASTA (Association of Short Circuit Testing Authorities)

This mark indicates that a product conforms to a National Standard. It is also associated with BEAB.

Atom

Atoms are the basic units of matter and comprise electrically positive (+ve) protons and electrically neutral neutrons that form a dense nucleus which is surrounded by a cloud of electrically negative (–ve) electrons.

There are 118 atoms, the first 88 of which occur naturally.

The simplest atom is that of hydrogen which has 1 proton and 1 electron.

Copper, used so frequently for cables, has 29 protons and 29 electrons.

Authorized person

This is usually a person who has demonstrated a specific level of competence within an organization which will allow him/her to switch/isolate and/or issue permits-to-work for low and/or high voltage systems.

Automatic disconnection of supply (ADS)

This is a means of providing protection against the risk of electric shock by

1. Basic protection (insulation of live parts, barriers or enclosures) and
2. Fault protection (earthing, bonding and the use of fuses, circuit breakers and RCDs)

Apart from earthing and bonding, ADS requires protective devices to operate within specified times and BS 7671:2008 provides tables of maximum values of loop impedance which will satisfy these disconnection times.

For TN systems from 120 V to 230 V a.c.

(a) all final circuits up to 32 A must disconnect in 0.4 s and

(b) final circuits above 32 A and distribution circuits must disconnect in 5 s.

For TT systems from 120 V to 230 V a.c. the times for (a) and (b) are 0.2 s and 1 s respectively.

For 110 V reduced voltage systems the disconnection time must not exceed 5 s.

There are three other methods of shock protection: double or reinforced insulation, electrical separation and SELV or PELV.

However, ADS applies to the majority of all complete installations; the others, generally, apply to specific circuits/equipment.

Autotransformer

This is a transformer with a single winding, the secondary being 'tapped' off the primary. They are used in the high voltage transmission system, and also generally as a means of providing the correct voltage to machinery, etc. (Fig. 2)

They may be 'step-up' or 'step-down' and also variable if required (variac).

The IET Wiring Regulations require that:

- If an autotransformer is used in a circuit with a neutral conductor, the common point on the winding should be connected to that conductor
- Where the transformer is a 'step-up' type, the disconnection of *all* live conductors must be achieved by a linked switch.

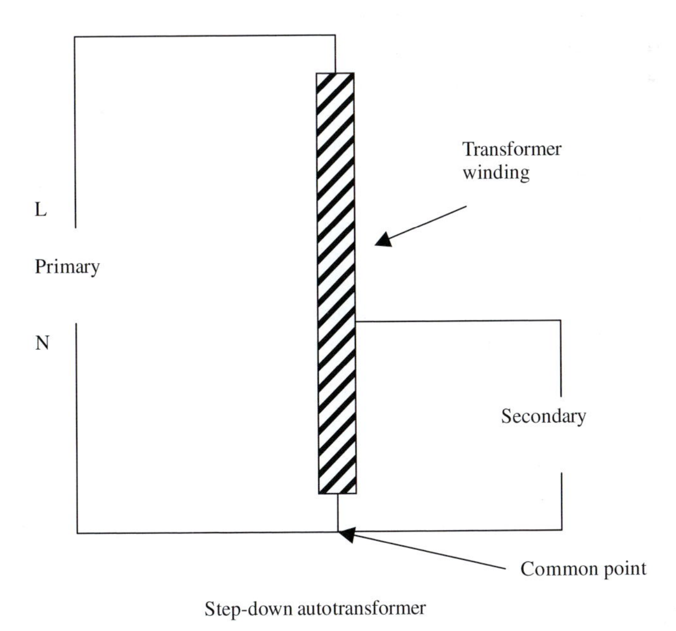

Step-down autotransformer

Fig. 2

B

Back e.m.f. (electromotive force)

When an alternating current flows in a circuit or item of equipment it produces an alternating magnetic field. This field changes direction 50 times a second, and as it does so the lines of force cut across the conductors in the circuit or equipment inducing e.m.f.s in them.

These e.m.f.s oppose the current that produces them and hence are in opposition to the flow of current. This opposition is known as inductive reactance, X_L, and is measured in ohms (Ω).

Back-up protection

This is used where a protective device is installed in a circuit and it has a lower breaking capacity than the prospective fault current at the point at which it is installed, but cannot be up-rated because it achieves discrimination ('catch-22' situation!).

Back-up protection should not be confused with additional protection by RCDs.

The 'back-up' device is placed in series with, and 'up-stream' (nearer the origin) of, the circuit protective device. Its purpose is to limit the 'let-through' energy during a fault.

The design of circuits requiring 'back-up' protection is complex, and the correct choice of devices is not easily accomplished.

Such situations are likely to arise in industrial locations, or where the supply transformers are close to the intake position of installations.

(see also Discrimination and Let-through energy)

The Dictionary of Electrical Installation Work. DOI: 10.1016/B978-0-08-096937-4.00002-7

Band I

This is the voltage band that normally encompasses extra-low voltage used for shock protection or operational reasons such as telecoms, bell, control and alarm installations *(see also Voltage bands)*.

Band II

This is the voltage band that normally encompasses low voltage used for supplies to household, commercial and industrial installations *(see also Voltage bands)*.

Barrier (see also Enclosure)

(BS 7671:2008 definition) 'A part providing a defined degree of protection against contact with live parts from any usual direction of access.'

Typical of this is the shield over the open bus-bar at the bottom of the protective devices in a consumer unit, or the internal cover plate behind the door of a distribution board.

The defined degree of protection would be the relevant IP code – for example IP2X or IPXXB as a minimum and IP4X or IPXXD as a minimum – for accessible horizontal top surfaces *(see IP codes)*.

BASEC

British Approvals Service for Cables. This is similar to BEAB. In this case it is cable that is subject to safety testing *(see also BEAB)*.

Basic insulation

This is insulation such as pvc, rubber, magnesium oxide, etc. which covers **live** parts and which can only be removed by destruction. It is intended to provide basic protection.

It is not to be confused with insulating material covering basic insulation. Such covering is called sheathing.

Basic protection

(BS 7671:2008 definition) 'Protection against shock under fault free conditions.'

This protects against the risk of shock from contact with parts that are intentionally live (direct contact) and is provided by:

1. Basic insulation, or
2. Barriers or enclosures.

Bathrooms

(BS 7671:2008 Section 701) These are locations that contain bath-tubs and showers with or without basins. So, they would apply to dwellings, sports facilities, leisure centres, etc.

The locations, which are divided into three zones, 0, 1 and 2, are as shown in Figs 3a, 3b and 3c.

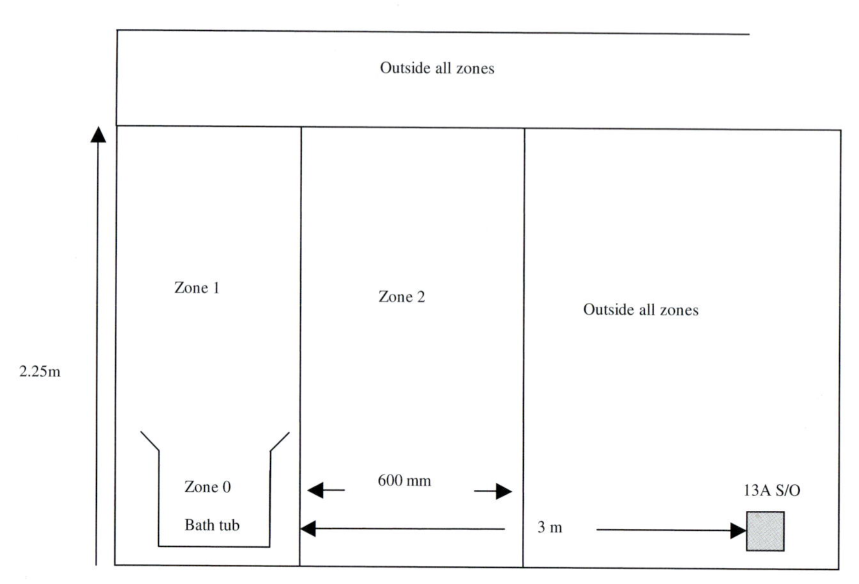

Fig. 3a

Main points:

- The space under the bath tub or shower basin is outside all the zones if that space can only be accessed by the use of a tool, for example to remove a surround panel. Otherwise it is part of zone 1

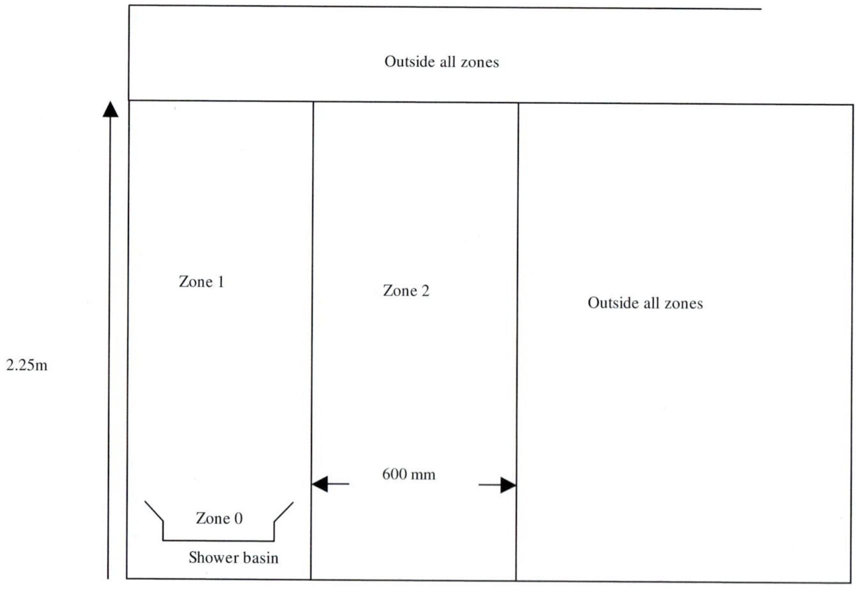

Fig. 3b

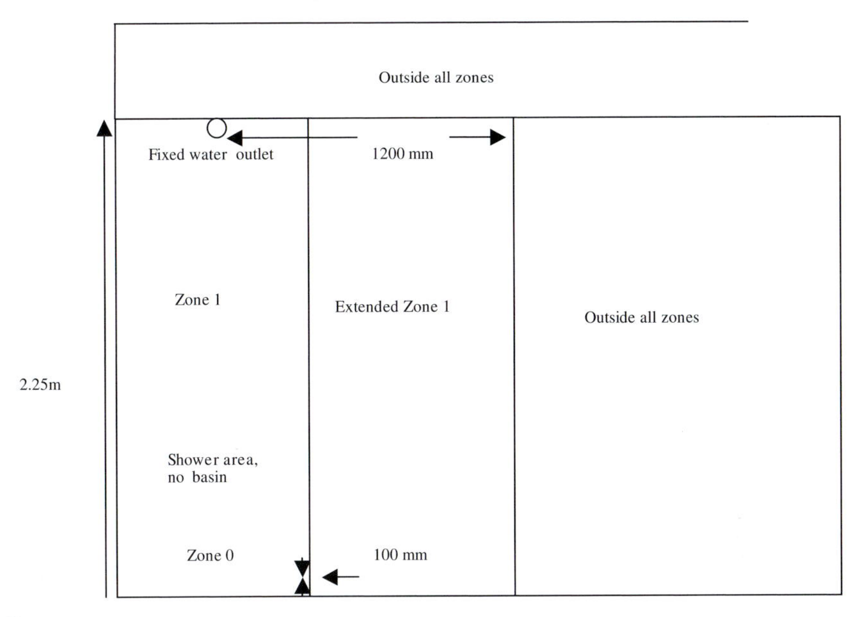

Fig. 3c

- There is no zone 2 for showers without basins, e.g. wet rooms, just an extended zone 1 which extends 1.2 m from the fixed water outlet on the wall or ceiling (no account is taken of demountable shower heads)
- 13A socket outlets may be installed beyond 3 m from the boundary of zone 1 (see Fig. 3a)
- All low voltage circuits of the location must have additional protection by an RCD rated 30mA or less
- Supplementary equipotential bonding is required connecting together the terminals of the protective conductors of Class I and Class II equipment to accessible extraneous conductive parts.

However, if all the final circuits are protected by automatic disconnection of supply *(almost certainly)*, all circuits are RCD protected *(a requirement anyway)* and extraneous conductive parts are effectively connected to the protective equipotential bonding system *(which is most likely if the main bonding has been carried out)*, then **no** supplementary equipotential bonding is required.

BEAB

British Electrotechnical Approvals Board. This is the UK National Certification Body for domestic and light commercial electrical equipment. A BEAB mark indicates that a product has been subjected to intensive and rigorous testing to ensure its safety *(see also ASTA)*.

Block diagrams (see Diagrams)

Bonding (see Equipotential bonding)

Breaking capacity

This is the value of fault current that a protective device can break and, in the case of a circuit breaker, without damage to itself or surrounding materials. It is usually quoted in kA *(see also Fuses, Circuit breakers and Prospective fault current)*. Protective devices must be able to operate effectively and safely at the value of prospective fault current at the point they are installed.

BS

Stands for British Standard. There are thousands of these, ranging from BS 2 Tramway and dock rails and fishplates (through to BS 61535-200 Installation couplers for permanent connection in fixed installations).

BS 7671:2008

This is the 17th edition of the IET Wiring Regulations and is a non-statutory document.

BS EN

This is a British Standard European Norm, e.g. BS EN 60898 for circuit breakers.

BSI

This is the British Standards Institution, which provides some 27000 standards globally. They also provide assessment and certification services, together with testing product quality and training services. It is also an approval body for Part 'P' of the Building Regulations.

Building Regulations 2000

The Building Regulations 2000 (statutory) comprise 14 Parts, those most relevant to electrical contracting being Parts A, B, E, F, L, M and P. For each of the Parts there is an Approved Document (non-statutory) which gives guidance on the means to achieve compliance with its associated Part *(see also Part 'P')*.

Bus-bar

An omnibus was the original term for a vehicle that carried many people. In the early days of the 'new technology of electricity' an 'omnibus bar' was a copper rod that carried the whole of the current of an installation.

Since those early days it has been abbreviated to 'bus-bar', and bus-bars are found, usually in larger installations, enclosed in housings known as bus-bar chambers or in bus-bar trunking systems. Bus-bars provide a facility to 'tap-off' in order to feed separate circuits or items of equipment. However, on a smaller scale, the copper strip connecting the bottom of protective devices in a consumer unit is a bus-bar.

C

C_a, C_g, C_f, etc (see Rating factors)

Cables

(BS 7671:2008 Appendix 4) Cables in the electrical contracting industry comprise one or more copper or aluminium conductors, each surrounded by insulating material, which is usually pvc or rubber or, in the case of mineral insulated (m.i.) cables, magnesium oxide.

Cable insulation is protected from mechanical damage by sheathing, armouring, copper cladding for m.i. cables, or enclosing non-sheathed single core cables in conduit, trunking, ducting, etc.

The assembly of cables, their enclosures and supports, etc is a 'wiring system' or 'cable management system'.

Appendix 4 of BS 7671:2008: gives details of various ways of installing cables. These are methods A, B, C, D, E, F and G and are as follows:
Method A.... Multi-core cables or non-sheathed and multi-core cables in conduit, where the cable or conduit is in contact with thermal insulation on one side only or where they are run in window frames or architraves. Also non-sheathed cables in mouldings.
Method B....Generally, all the standard cable types enclosed in conduit, trunking, ducting, floor channel, building voids, etc where thermal insulation is not present.
Method C....Sheathed single-core and multi-core cables mounted direct to a surface or un-perforated tray or buried in non-thermal masonry or plaster. This method is usually referred to as 'clipped direct'.
Method D....Non-armoured single or muilti-core cable in conduit or ducts underground. Sheathed, armoured or multi-core cables direct in the ground.
Method E or F....Single-core or multi-core cables on perforated tray or brackets or ladders, etc.

The Dictionary of Electrical Installation Work. DOI: 10.1016/B978-0-08-096937-4.00003-9

Method G....Non-armoured cables on insulators, e.g. overhead lines.

(See also Current carrying capacity and Design current)

Cables in walls or partitions (see also Residual current device)

BS 7671:2008 requires that protection be given to cables in walls or partitions from the effects of shock caused by penetration by nails and screws, etc.

Calibration of test equipment

The Electricity at Work Regulations 1989 require electrical systems to be regularly maintained in order to avoid danger. The words test, inspection, calibration, etc are not mentioned but are implicit in the word 'maintained'.

Items of test equipment are 'systems' and should be maintained in a safe condition and hence regular checks on their condition and accuracy are required.

It is recommended that accuracy is confirmed and recorded at regular intervals by checking against known values ('check-boxes' are commercially available).

A comparison of accuracy against National Standards is recommended, although it is not mandatory, every year or at such intervals as is deemed necessary dependent on the frequency of use of the equipment.

Candela cd

This is best described as the unit of brightness of a light source. Once called candle power.

Capacitance C farads

This is the property of a circuit or component to store electrical energy.

Capacitive reactance X_C ohms

This is an opposition, caused by capacitance, to current in an a.c. circuit.

Capacitor

A component that stores electrical energy for a short period of time and found in installation work in such items as single phase motors for starting purposes, fluorescent luminaire starters for radio interference suppression, discharge lighting and whole installations (usually large industrial) for power factor correction.

Caravans

(BS 7671:2008 Section 721) Those that are used for habitation, i.e. the types that are in static positions are **not** included in special locations. They include those that are towed or are motor caravans.

Main points:

- Where protection is by automatic disconnection of supply, a double pole RCD of maximum rating 30 mA shall be provided
- Periodic inspection and testing shall be carried out, preferably not less than once every three years, or once a year for frequently used caravans

- Inlets should be BS EN 60309-1 or 2, and be 1.8 m above ground level and rated IP44
- Supply cables should be 25 m (+/– 2 m) long
- Cable plugs for connecting to the pitch supply should be to BS EN 60309-2.

Caravan and camping parks

(BS 7671:2008 Section 708) These are the areas for supplying electrical energy to caravans and tents.

Main points:

- Electrical equipment should withstand the external influences of water, foreign solid bodies and impact by ensuring it is coded at least IPX4, IP3X and IK08 respectively
- Overhead cables should be 6 m above ground level in vehicle movement areas and 3.5 m in all others, and support poles be placed to avoid damage
- Underground cables should be at a depth of at least 600 mm and, if without additional protection, they should be be as far outside the caravan pitch as possible in order to avoid tent pegs, etc
- Socket outlets should be:
  - to BS EN 60309-2, not less than 16 A and at least IP 44 rated
  - a maximum of 4 per pitch
  - individually protected against overcurrent
  - individually protected by a 30 mA or less RCD
  - between 0.5 m to 1.5 m from ground level to the bottom of the outlet. This height may be exceeded in circumstances where there is a risk of flooding or heavy snowfall.

For PME (protective multiple earthing) systems the protective conductor of each socket outlet must be connected to an earth electrode, thus converting it to a TT system.

CDM

The Construction (Design and Management) Regulations 1994.

This requires architects, designers and managers to formulate a safety policy for a particular project.

CE mark

The CE mark is an indication by the manufacturer or importer of goods into the European Union that a product complies with the EMC (electromagnetic compatibility) and the LV (low voltage) Directives.

This marking is **not** an indication of product quality and, in the end, wholesalers, contractors and end users will still have to ensure that products are reliable, robust and safe, and that they come from reputable manufacturers.

CENELEC

Comité Européen de Normalisation Electrotechnique or European Committee for Electrotechnical Standardization.

This body is responsible for the standardization of electrical engineering in Europe.

Certification

(BS 7671:2008 Appendix 6) There are three certificates and relevant documents that may be completed depending on the work carried out, these are:

1. An Electrical Installation Certificate (EIC) for new installations or alterations or additions.
2. A Minor Electrical Installation Works Certificate (MEIWC) for alterations or additions that do not include a new circuit.
3. An Electrical Installation Condition Report (EICR) for reporting on the condition of an existing installation.

An EIC and an EICR **must** be accompanied by schedules of test results and inspections. Without them the certificates are invalid.

EICs and MEIWCs are signed or otherwise authenticated by the person/s responsible for the design, the construction and the inspection and testing of the installation.

EICRs are signed or otherwise authenticated by the person/s carrying out the inspection and testing of the installation.

CFL lamp

Compact fluorescent lamp. These are energy saving lamps.

Circuit diagrams (see Diagrams)

Circuits

In electrical installation work, there are two main types of circuit:

- The ring and
- The radial

The rings can be further divided, typically, into

- Ring final circuits (for socket outlets)
- Overhead bus-bar trunking rings with an isolator at the mid point to allow half the ring to be isolated at a time for maintenance purposes.

The radials may be

- Distribution circuits, or
- Final circuits for lighting, power, etc.

Circuit breakers (see also RCBOs)

These are electro-mechanical protective devices capable of making, carrying and breaking normal load currents and automatically breaking or manually making overcurrents. They are usually referred to as miniature circuit breakers, MCBs, although BS 7671:2008 refers to them as 'circuit breakers' and does not use an abbreviation.

They comprise two parts:

1. A thermal (bi-metal strip) element that protects against overloads, and
2. A magnetic solenoid that acts instantaneously to protect against fault currents.

BS EN 60898 circuit breakers are the most common and are available in types B, C and D.

Type Bs have characteristics that can allow an overload of up to 5 times their rating and hence are suited to installations where overloads are generally unlikely, such as domestic installations, and small shops and offices.

Type Cs can allow up to 10 times their rating and are suitable for light industrial and large commercial applications.

Type Ds can allow up to 20 times their rating, and they are likely to be found in heavy industrial locations where there may be substantial motor starting currents or inductive loads, and medical environments where X-ray machines are present.

Circuit breaker specifications quote two breaking capacities: I_{cn} which is the maximum current that it can interrupt safely (it may not be functional after this level), and I_{cs} which is the level it can interrupt safely and remain effective. The I_{cn} kA value is normally shown on the breaker, e.g. 10000

For values up to 6 kA the I_{cn} and I_{cs} values are the same.

It should be noted that whilst the current ratings of circuit breakers are the same for each of the types, the maximum loop impedance values for a type C is smaller than for a type B, and a type D smaller than a type C.

So, although, for example a 20 A type C may be suitable for the installation application, its value of loop impedance may prohibit its use because of the risk of shock!

BS 3871 miniature circuit breakers, although obsolete, are abundant in older installations, and, whilst not considered unsafe for continued use, those needed for use in spare ways of a distribution board should be replaced with BS EN 60898 types. Most manufacturers make BS EN 60898s which are dimensionally equivalent to BS 3871s and hence will fit in older boards.

Within the classification of circuit breakers are MCCBs (moulded case circuit breakers) which perform the same function as circuit breakers but are more suitable for applications where high breaking capacity and speed of operation is important.

Circuit protective conductor (cpc)

This is the conductor/s that connects exposed conductive parts of equipment to the main earthing terminal (MET) of an installation.

Such a conductor need not necessarily be a single core cable or a core in a cable; it could be the metal sheath or armour of a cable, or metal conduit or trunking, etc., or even, in special circumstances, an exposed conductive part itself. It is not usual, however, to find modern installations using conduit or trunking as a cpc.

A cpc provides part of the measure used for fault protection by 'automatic disconnection of supply' ADS.

The size of a cpc may be selected from the BS 7671:2008 table 54.7 or calculated by using the adiabatic equation:

$$S = \frac{\sqrt{I^2 \cdot t}}{k}$$

Where S is the conductor size

I is the fault current

k is a factor dependent on the conductor materials.

Circuses (see Fairgrounds)

Class I equipment

This is equipment that is not only reliant on basic insulation, but also requires the provision of a connection to earth for shock protection. Basically, such equipment, metal-cased or not, has a cpc in its supply cable.

Class II equipment

This equipment relies on basic insulation plus supplementary or reinforced insulation to provide shock protection, so it does not need any protective conductor.

It is usually referred to as double insulated equipment and is symbolized ⧈

Class III equipment

This is equipment that is supplied from a SELV source, and is typical of office equipment such as fax machines, telephones, etc. or lighting, jacuzzis; etc. in some modern bath-tubs.

It is symbolized

Concentric cable

This is a single or three core cable surrounded by armouring, which is normally copper. The armour provides the function of both earth and neutral i.e. a PEN conductor. *(Used on TN-C-S systems.)*

Another version of this arrangement is where half the armour is sheathed and the other half bare. This is called **split concentric**. *(Used on TN-S systems.)*

Co-axial cable used for TVs etc is a concentric cable.

Conducting locations with restricted movement

(BS 7671:2008 Section 704) Such locations are uncommon. They comprise metallic surrounding parts, such as large ventilation ducting or pressure vessels, within which a person's movement is severely restricted.

Main Points:

- Supplies for hand-tools etc. can be protected by
  - electrical separation, or
  - SELV
- Supplies for hand lamps
  - SELV
- Supplies for fixed equipment
  - ADS with supplementary equipotential bonding, or
  - Class II equipment with additional protection by 30 mA or less RCDs
  - Electrical separation, or
  - SELV, or
  - PELV with extra bonding inside the location ans the PELV connected to earth.

Conductivity

This is the ability of a material to conduct electricity.

Conduit

A conduit is an enclosure or containment system that is used to minimize the risk of damage to cables. It may comprise a complete rigid system or isolated rigid lengths for cable drops to accessories, or flexible types for connection to equipment.

The most common rigid types are heavy duty, black enamelled or galvanized welded steel or standard or heavy duty pvc. The most common sizes are 20 mm or 25 mm diameter, with standard lengths of 3.75 m for metal and 3 m for pvc.

Flexible conduit may be metal, pvc covered metal, nylon or polypropylene and it is available in ranges from 16 mm to 33 mm dia. for metal and up to 57 mm for non-metallic.

Steel conduit may be used as a cpc although rarely in modern installations, where a separate cpc is provided. The 'fly-lead' used to connect an accessory to a 'back-box' is only necessary where the conduit is used as the cpc.

Flexible metal conduit must **not** be used as a cpc.

Oval pvc conduit is often used for cable drops to accessories. The use of such conduit is not intended for cable withdrawal, or the containment of single core cables, or mechanical protection against nails, screws, etc. It is just a protection for cables from damage by the plasterer's trowel during the 'first fix' stage of an installation. The same is the case for metal or pvc 'top-hat' sections.

However, metal conduit when embedded in walls will provide mechanical protection against penetration by nails, screws and the like.

Pvc conduit is most suited to light duty applications and where there is a risk of corrosion.

Most pvc conduit is manufactured to be rodent proof.

Conduit capacity

In order to facilitate the ease of 'drawing-in', a limit is placed on the number of cables permitted. This number is dependent on the size of the conduit, the length, and the number of bends or sets within a conduit.

The IET On-Site-Guide gives **guidance** on this in a tabulated form. The figures in the tables may need adjusting to take account of grouping and varying thickness of cable insulation.

In the absence of tabulated values, a 'space factor' of 40% can be applied. This simply means that cables should only occupy 40% of the space in the conduit.

(see also Trunking capacity)

Construction and demolition sites

(BS 7671:2008 Section 704) This section deals with construction, alterations, repairs, demolition, earthworks, etc. It does not cover site offices, toilets, canteens, dormitories, etc.

Main Points:

- Socket outlet circuits up to 32 A and other circuits feeding hand held equipment up to 32 A may be protected by:
  1. Reduced low voltage (110 V CTE), or
  2. ADS with additional protection by 30 mA or less RCDs, or
  3. Electrical separation, or
  4. SELV or PELV.
- Option 1 is preferred for hand held equipment, lamps and tools up to 2 kW
- Option 4 is preferred for handlamps in damp confined spaces
- Socket outlets exceeding 32 A rating shall be protected from dead shorts between lines to exposed conductive parts by an RCD rated not more than 500 mA and which automatically interrupts the supply to all line conductors
- Cables crossing site roads or walkways must be protected against mechanical damage
- Site supplies should be fed from an Assembly for Construction Sites (ACS) comprising fault and overcurrent protective devices and socket outlets, if required
- The Electricity, Safety, Quality and Continuity Regulations 2002 (ESQCR) prohibits a PME system on a construction site, except for the supply to a fixed building of the site.

Construction Skills Certification Scheme (CSCS)

This organization was set up to help improve health and safety in the workplace. An operative may apply for and obtain a CSCS card, which is an indication of occupational competence. The Electrical Certification Scheme (ECS) card is affiliated to the CSCS and applies to electrical operatives. It is administered by the Joint Industry Board (JIB).

Contactor

An item of manual or automatic equipment used to control, for example, heating/lighting systems or motors. They may be either single or three phase.

(see also Starter and Hold-on circuit)

Continuity (see Testing)

Copper losses

Also known as I^2R losses, these are power losses in conductors, particularly within transformers, inductors and motors due to the extreme length of those conductors.

(see also Hysteresis)

Corrosion

This may occur wherever corrosive substances are present, such as salt water, chemicals, hydrocarbons, etc., or where dissimilar metals are in close proximity in wet or damp environments.

Corrosion may occur, for example, where steel wire armoured cable is installed outdoors. The steel termination is via a brass gland and hence a pvc or rubber shroud is placed over the termination to protect the dissimilar metal join from moisture. It is usual, however, to use a shroud in any event, regardless of the environment, for aesthetic reasons.

COSHH

This stands for the Control of Substances Hazardous to Health Regulations 2002.

They require an assessment of the risks of, and the appropriate actions needed, with regards to hazardous substances.

CSCS (See Construction Skills Certification Scheme)

Current (I amperes)

This is the flow of electrons in a circuit. The actual electron flow is from negative (–ve) to positive (+ve). It was originally thought that electric current was the flow of protons from +ve to –ve. However, as there are the same number of electrons as there are protons in an atom, the convention of current flowing from +ve to –ve has been left and is known as conventional current flow.

Current-carrying capacity of a cable (Iz)

This is the maximum current that a cable can carry, safely, under the conditions in which it is installed. For example, a cable may have a current rating of 20 A but, due to adverse conditions along its route, it may have to be de-rated to carry only 15 A so that it does not overheat. This latter value is I_z.

D

d.c. (direct current)

This is usually produced by batteries, but it can be derived from d.c generators, or electronically from a.c to d.c rectifiers.

Delta connection

This is one way that three phase supplies or loads may be arranged. It is not usual for a standard low velocity (LV) supply to be delta as there would be no neutral, the delta arrangement is on the high velocity (HV) side of the supply transformer.

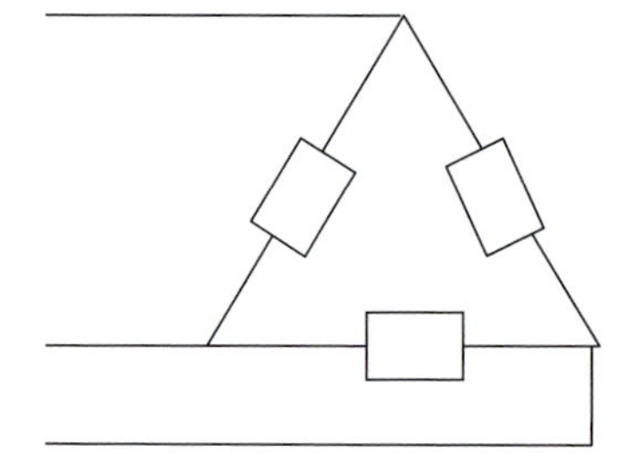

Delta connection of windings or loads

Fig. 4

Three phase motors are generally delta connected, as their windings are all the same and hence a neutral is not required. For large motors with heavy starting currents, their windings are Star connected at start-up and then automatically changed to delta when they reach a suitable speed *(see also Star connection)*.

Design current (Ib)

(BS 7671:2008 definition) The magnitude of the current (rms for a.c.) to be carried by the circuit in normal service. This does not include, for example, inrush currents caused by motor starting or switching inductive loads such as discharge lighting ballasts, etc. Design current may be determined from manufacturers' information or calculated from:

$$\text{Single Phase.......Ib} = \frac{\text{power in watts}}{\text{V (usually 230)} \times \text{pf} \times \text{Eff\%}}$$

$$\text{Three Phase.......Ib} = \frac{\text{power in watts}}{\sqrt{3} \times V_L \text{(usually 400)} \times \text{pf} \times \text{Eff\%}}$$

The Dictionary of Electrical Installation Work. DOI: 10.1016/B978-0-08-096937-4.00004-0

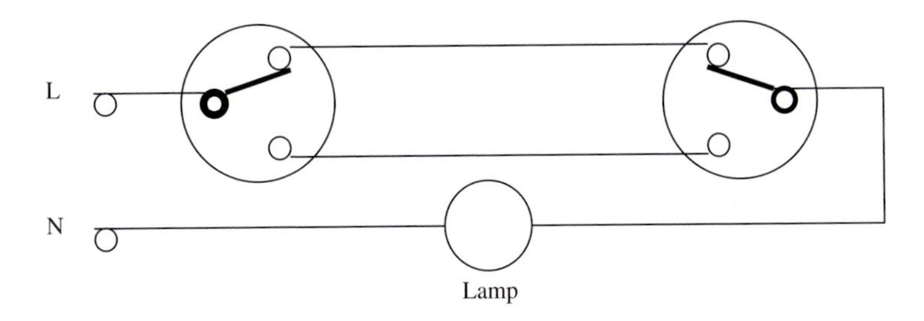

Fig. 5

Where pf = power factor if needed

Eff% = efficiency if needed

Diagrams

There are various diagrams associated with electrical installations, the most common are:

- The CIRCUIT or SCHEMATIC diagram, which shows how a system ***works***, and not how it would actually be wired. Such a diagram, Fig. 5 for example, shows a two way lighting system
- The WIRING diagram, which indicates how a circuit is actually ***wired***
- The BLOCK or LAYOUT diagram, which shows generally how items of equipment are connected. A example of a three-phase system is shown in Fig. 6.

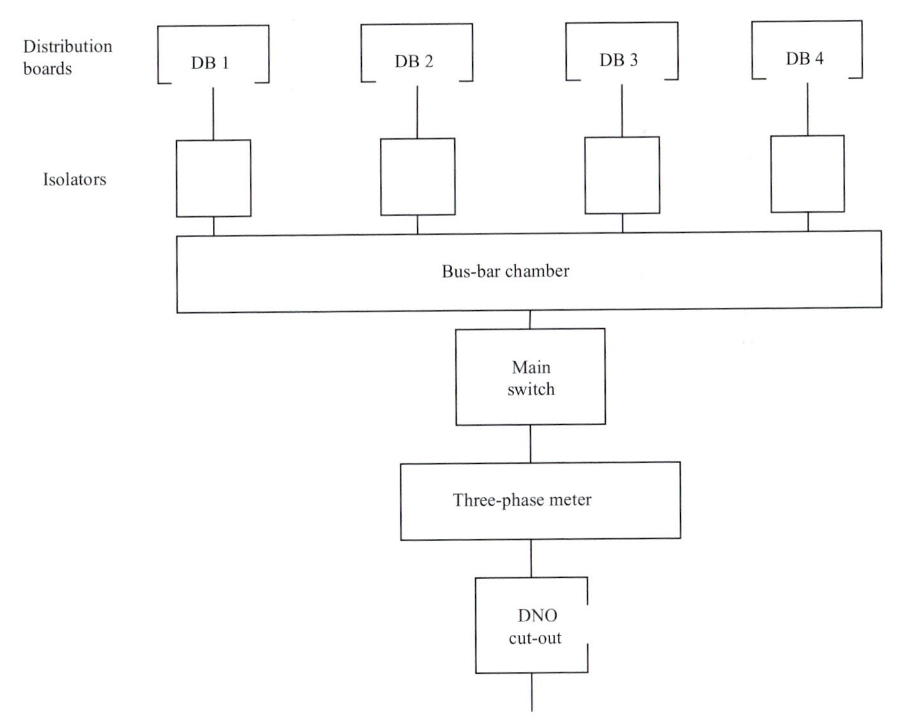

Fig. 6

- The INTERCONNECTION diagram is similar to the block or layout diagram but has more technical detail regarding size of cables, rating of equipment, etc.
- The ARCHITECTURAL diagram, which indicates the positioning of equipment and accessories and the routes of cables on drawings and plans. The symbols used on such drawings should be to BS EN 60617, examples of which are shown below.

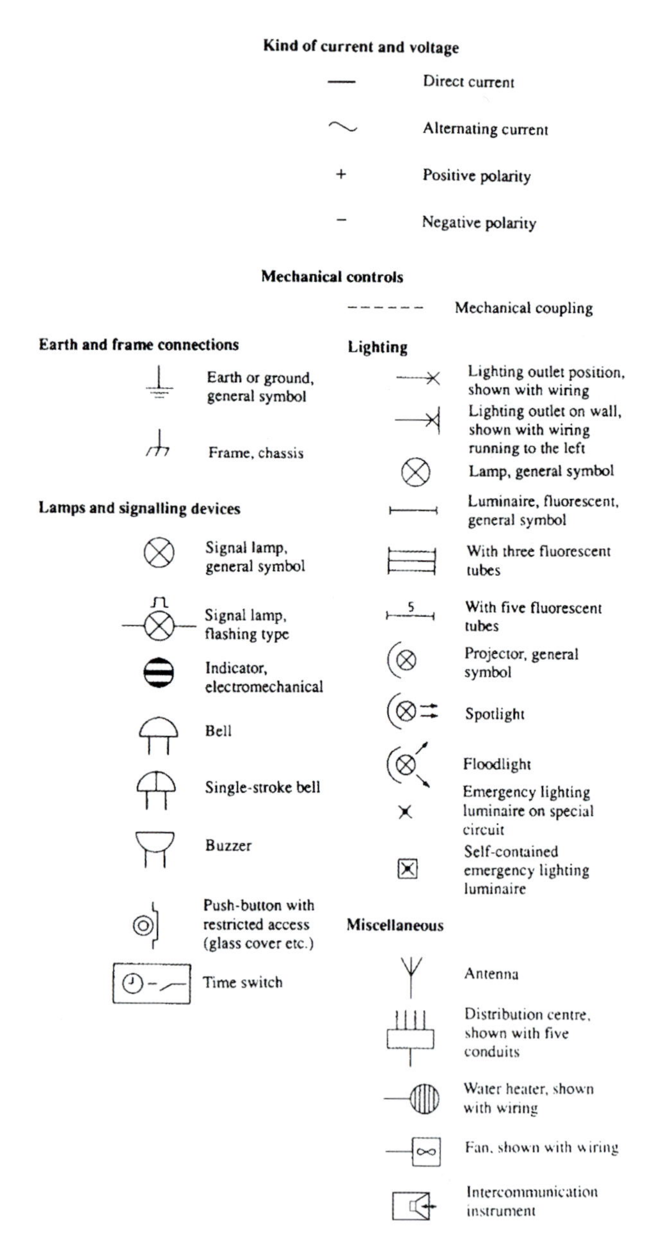

continued on the next page

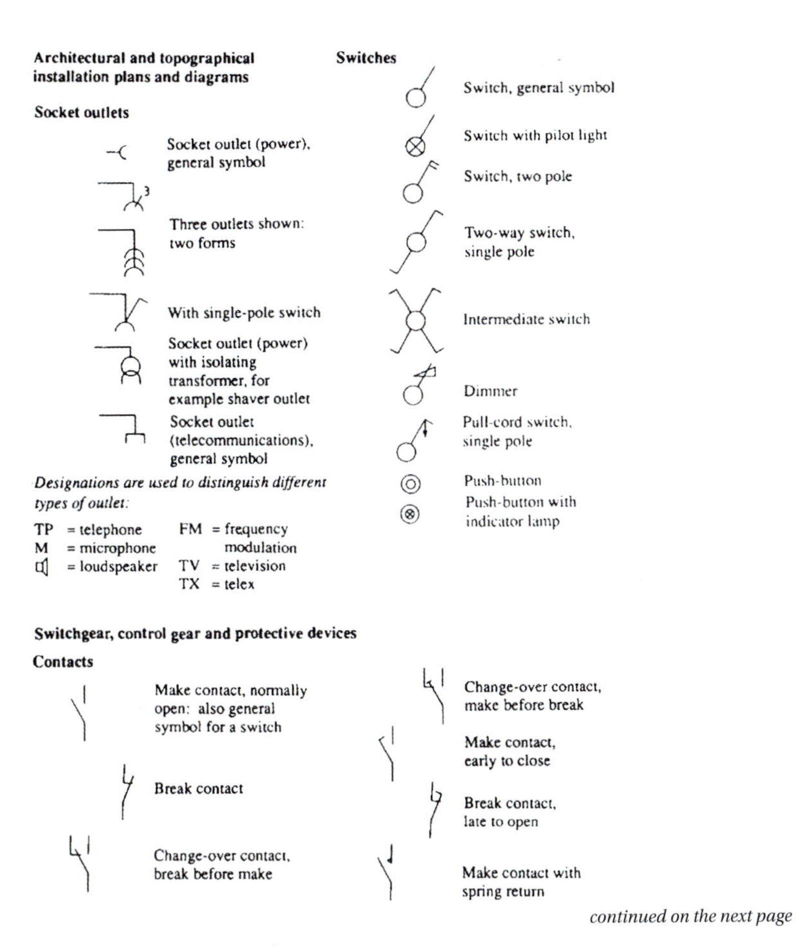

continued on the next page

Direct contact (see Basic protection)

Discharge lighting

Included in this range are:

Low pressure mercury vapour (fluorescent) ...*General purpose lighting*
High pressure mercury vapour ... *Street lighting*
High and low pressure sodium vapour ... *Street lighting; high and low bay lighting*
Metal halide ... *High intensity illumination, used extensively in horticulture*
High voltage lighting ... *e.g. coloured shop signs and displays*

This type of lighting requires starting arrangements that include either transformers or inductors (ballasts/chokes), which can cause power factor issues. Hence, when determining the volt-ampere rating of such lighting, and in the absence of technical details, the lamp watts are multiplied by a figure of not less than 1.8.

Hence a discharge luminaire housing a 300W lamp would be rated at
$300 \times 1.8 = 540$VA

(see also Power factor)

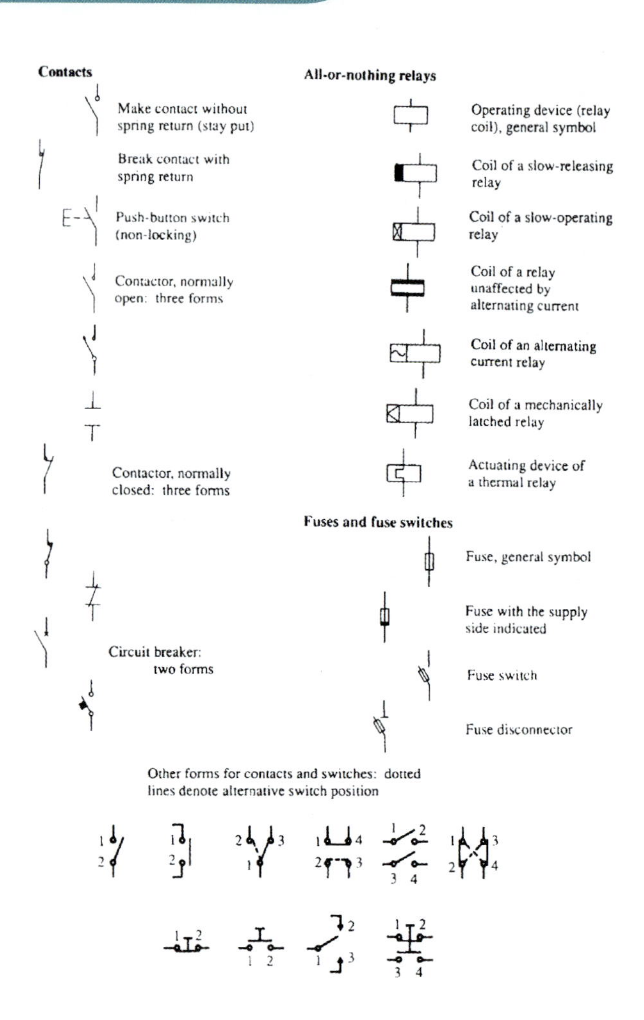

Fig. 7

Disconnection times (see Automatic disconnection of supply)

Disconnector (see Isolator)

Discrimination

This is required to ensure that the correct protective device operates in the event of a fault, i.e. the minor device should operate before the major device.

For example a fault on, say, a lighting circuit protected by a 6A circuit breaker should not cause the main service fuse to operate!

Discrimination between devices is achieved if the **total** let-through energy (I^2t) of the minor device is less than the pre-arcing let-through energy (I^2t) of the major device. Usually, one size difference will achieve discrimination, but this cannot always be assured and reference to manufactures' information is important.

RCDs, being such sensitive devices, may not provide discrimination and where it is important, time delay or 'S' types may need to be employed.

Distribution circuit

This is a circuit that supplies switchgear, or distribution boards, or outlying buildings. In the latter case they are often referred to as sub-mains.

The tails from the supply to a consumer unit is a distribution circuit.

Distribution Network Operator (DNO)

These are the organizations that deliver electrical energy to installations. It is the generic term for the Regional Electricity Companies, so, SWEB, MANWEB, SEEBOARD, etc are all DNOs.

Diversity

If the maximum demand of an installation were used to establish the rating of the main switchgear, distribution cables, service cables, metering, etc., then such equipment would, more than likely, be grossly oversized.

For example, in the case of the electrical installation in a standard three bedroom premises comprising 2..32A ring final circuits, 2..6A lighting circuits, 1..40A shower circuit, 1..16A immersion heater circuit and 1..32A cooker circuit, the total possible load would be in the region of 160A, which is clearly too high for standard intake equipment which is usually rated at 100A.

The application of diversity assumes that individual circuits are unlikely to be fully energized, and that all circuits are also unlikely to all be energized at the same time. This assumption will significantly reduce the maximum demand to a more realistic level.

Suggested diversity values are given in The On-Site-Guide and Guidance Institute of Engineering and Technology's Note1 for small domestic and commercial installations. Larger or industrial type premises will need specialized knowledge to make decisions regarding diversity.

A useful means of demonstrating how diversity can reduce maximum demand is by considering a cooker circuit supplying, say, a cooker with a rated full load of 9.2kW.

Guidance notes suggest that the assumed current demand of a cooker is:

The first 10A of the connected load $+30\%$ of the remainder.

So our cooker would have a full load of $\frac{9200}{230} = 40\text{A}$

Hence, assumed demand would be 10 + 30% of $30 = 10 + 9 = 19\,\text{A}$ (a significant reduction).

A further 5A would be added if the cooker unit had a socket outlet.

This does not mean that the cooker final circuit should be de-rated for the purpose of cable sizing.

Double insulation (see Class II equipment)

Duct

This is a metallic or insulated enclosure other than conduit or trunking for the containment of cables.

Duty-holder

This is the title given, as per the Electricity at Work Regulations 1989, to anyone who has control of an electrical system. Control in this sense is the design, or installation, or inspecting/testing or maintaining or repairing, or even ownership of such a system.

E

Earth

(BS 7671:2008 definition) 'The conductive mass of earth, whose electric potential at any point is conventionally taken as zero.'

In other words it's the stuff we grow our spuds in and it's 0 volts !

Earth electrode

This is a conductive part that is imbedded in soil or in concrete, etc., and so is in contact with the earth.

An earth electrode may be any of the following:

- Earth rods or pipes
- Earth tapes or wires
- Underground structural metalwork in foundations
- Metal reinforcement of concrete imbedded in the earth
- Some lead sheaths or other metal coverings of cables
- A metal water pipe (not a utility pipe), provided precautions have been taken to avoid its removal and it has been considered for such use
- Other suitable underground metalwork.

The most common and familiar types are rods and plates or mats.

The following may **not** be used as an earth electrode:

- Metal gas pipes
- Metal pipes containing flammable liquids
- Metal water utility (service) pipes.

Earth electrode resistance

This is the resistance of the contact between an electrode and the surrounding earth.

(see also Testing)

The Dictionary of Electrical Installation Work. DOI: 10.1016/B978-0-08-096937-4.00005-2

E

Earth electrode resistance area

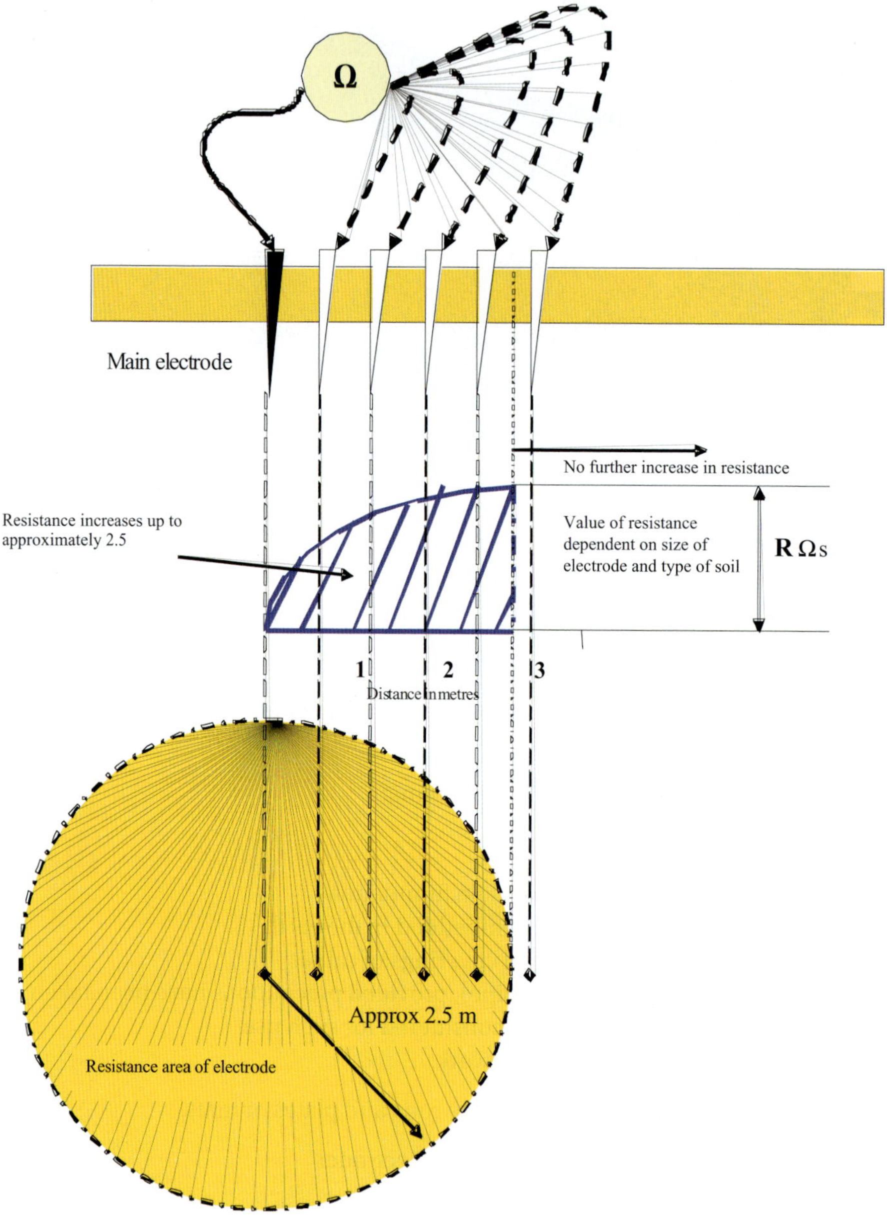

Ω

Main electrode

No further increase in resistance

Resistance increases up to approximately 2.5

Value of resistance dependent on size of electrode and type of soil

R Ωs

1 2 3

Distance in metres

Approx 2.5 m

Resistance area of electrode

Measurements taken at increasing distances from the main electrode would result in an increase in resistance out to approximately 2.5 m–3 m, after which there would be no further increase.

Fig. 8

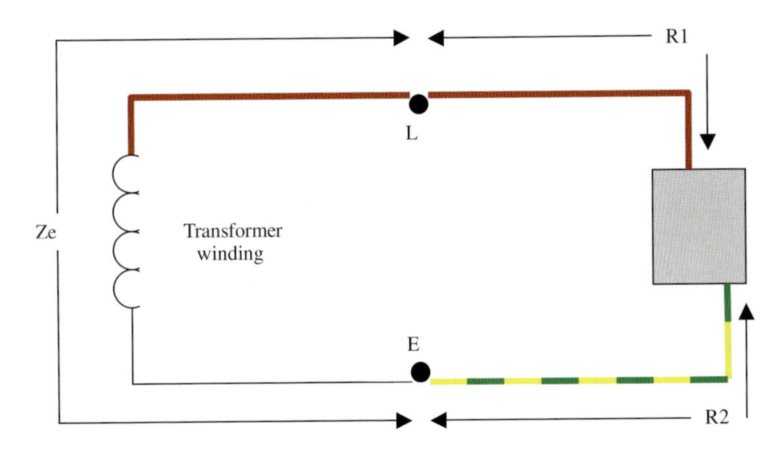

Fig. 9

Earth electrode resistance area

If a measurement between an electrode and the surrounding earth were taken at increasing distances from the electrode, it would be seen that the resistance values increased until a point was reached, about 2.5 m to 3 m from the electrode after which no further increase would be noticed (Fig. 8).

This circular area, of radius 2.5 – 3 m, is the resistance area of the electrode.

Earth fault current

(BS 7671:2008 definition) 'An overcurrent resulting from a fault of negligible impedance between a line conductor and an exposed-conductive-part or a protective conductor.'

Or, basically, a dead short between line and earth!

Earth fault loop impedance Zs (see also Earthing systems and Testing)

This is the resistance of the route that earth fault currents take, from the point of fault, through the internal (R1 + R2), and external (Ze) parts of the earthing system (Fig. 9).

Hence,

$Z_s = Z_e + (R1 + R2)$ ohms

Or more fully

$$Z_s = Z_e + \frac{\{(R1 + R2) \times \text{length} \times \text{multiplier}\}}{1000}$$

(see also R1 + R2 and Temperature coefficient)

Earth leakage (see Protective conductor current)

Earthing

(BS 7671:2008 definition) 'Connection of the exposed conductive parts of an installation to the main earthing terminal of that installation.'

This is achieved by the use of circuit protective conductors and should not be confused with bonding *(see also Equipotential bonding).*

The purpose of earthing is to limit the duration of voltages (sometimes called touch voltages) that may occur during an earth fault by disconnecting the protective device in the required time.

Earthing conductor

The earthing conductor is found at the origin of an installation and is often incorrectly referred to as the **main** earthing conductor.

It is a protective conductor that connects the main earthing terminal (MET) of an installation to the means of earthing such as an electrode or cable sheath or other mean of earthing.

The size of the earthing conductor may be determined by use of the Adiabatic equation or by selection from BS 7671:2008 Table 54.7

Earthing systems

The systems used for the UK public supply network are designated by a combination of the letters T, N, C and S.

The letter **T** denotes a direct connection to the mass of earth *(T is the first letter of the French word terre meaning earth)* and can refer to the supply source or the installation.

The letter **N** denotes a connection of the exposed conductive parts of an installation to a conductor provided by the DNO.

The letter **S** denotes separate metallic earth and neutral conductors.

The letter **C** denotes combined metallic earth and neutral conductors.

So, a TN system is a generic term for a system where the supply source has one or more points connected directly to earth and exposed conductive parts of the consumers' installation are connected to a conductor provided by the DNO. Such a system may be :
TN-C where the DNOs conductor performs the combined functions of both earth and neutral throughout the supplier's **and** the consumer's installations. Not very common: consumers' installation wiring difficult.
TN-S where the DNOs and the consumers provide a separate earth and neutral conductor throughout the whole system. Typical of a large percentage of the building stock in the UK.

TN-C-S where the DNOs supply of earth and neutral are combined in one conductor (PEN conductor) and the consumers' installation of earth and neutral are separate. The DNOs part is known as protective multiple earthing PME because their PEN conductor is earthed at many points along it length in an attempt to keep it at zero volts. PEN stands for **protective earthed neutral** *(see also Protective multiple earthing).*

It must be remembered that the TN-C-S system provides an artificial earth, as the neutral can and does carry current and, hence, the PEN conductor could have a potential above true earth. This can be problematical and the ESQCR prohibits the use of PME for a supply to a caravan or similar construction which would be found in some special locations The TN-C-S system is generally used for all new DNO supplies.

A TT system, is where the DNOs provide a source earth and the consumers provide their own earth. It is typical of overhead line supplies in rural areas.

BS 7671:2008 also lists the IT system in which only the consumers' installation is earthed. This system is not permitted for UK public supply systems, but may be encountered in special installations such as medical locations. In this case the letter **I** denotes that all live conductors are either isolated from earth or one point earthed through a high impedance.

(see also Insulation monitoring devices)

The following diagrams illustrate the TT and TN systems.

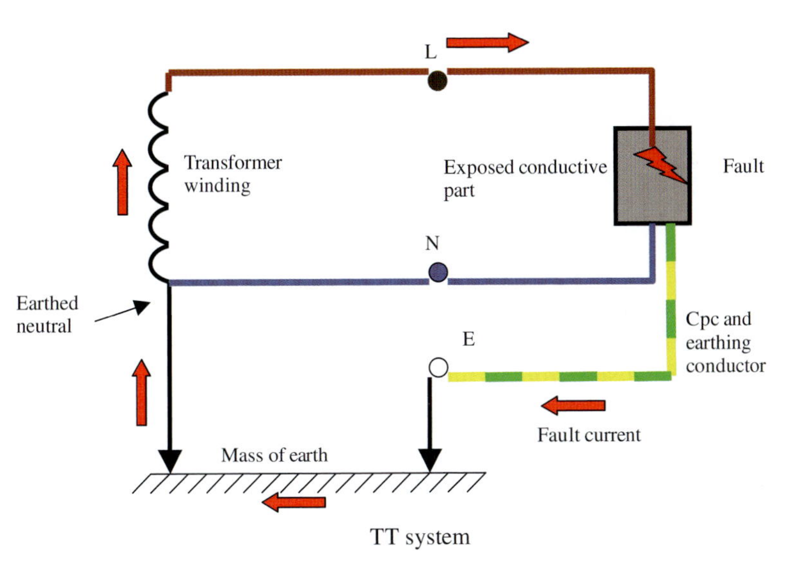

Fig. 10a

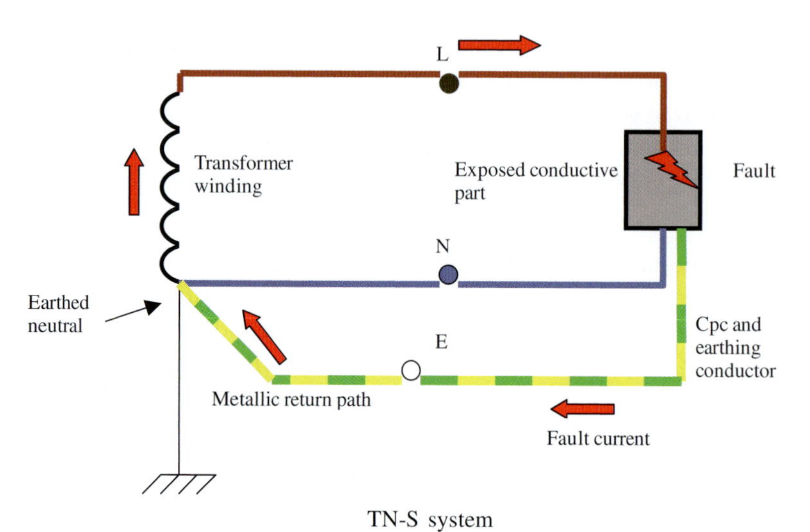

Fig. 10b

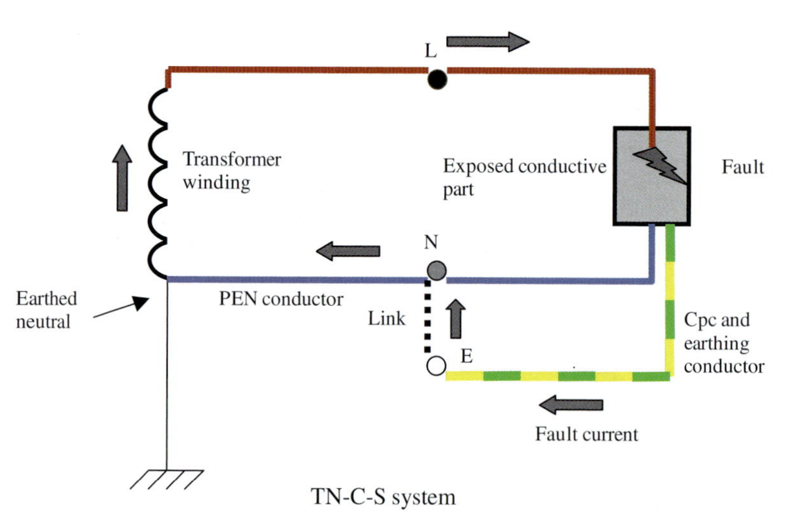

Fig. 10c

ECA

This is the Electrical Contractors Association, whose aim is to ensure a high quality of workmanship from its members. It also ensures that, should one of its member companies cease to trade, the customer is not left with uncompleted work. It is an approval body for Part 'P' of the Building Regulations.

ECS card (see Construction Skills Certification Scheme)

Eddy currents

These are small circulating currents induced in metals which are in close proximity to changing magnetic fields.

In large items, such as transformer cores and motor armatures etc., these eddy currents can combine and build up to create currents that are large enough to cause overheating problems. To overcome this, the cores are constructed of laminations which are insulated from each other. This prevents the circulating current build up.

Eddy currents can, however, be use to beneficial effect in non-ferrous materials, e.g. aluminium, due to the magnetic fields they produce. These fields may be used to dampen the speed of moving items thus providing a braking effect.

They are also used to good effect in recycling plants, to segregate aluminium cans etc. from other waste by 'throwing' the waste aluminium out from a centrifuge whilst subjecting it to magnetic fields. The eddy currents and associated magnetic fields induced in aluminium waste react with the main field causing the aluminium items to slow down and 'drop' out before other items.

(see also Ferromagnetic materials)

ELECSA

This is part of the ECA group. It is an approval body for Part 'P' of the Building Regulations.

Electric shock

(BS 7671:2008 definition) 'A dangerous physiological effect resulting from the passage of electric current passing through a human body or livestock.'

There are various levels of a.c. shock current that cause corresponding effects. These levels and effects cannot be firmly set and are likely to vary from person to person depending on health, age, etc. and the voltage present.

However, the following gives an indication of shock currents and resulting effects at around standard mains voltage. Note, the values given are in mA, i.e. ***thousandths of an ampere***

1 to 2 mA.............Barely perceptible. No harmful effects.
5 to 10 mA............Throw off. Painful sensation.
10 to 15 mA...........Muscular contraction, can't let go.
20 to 30 mA...........Impaired breathing. Asphyxiation starts.
50 mA and above.....Ventricular fibrillation. Cardiac arrest.

So, at only 1/20th of an ampere, death is very likely.

Electrical installation certificate EIC (see Certification)

Electrical installation condition report EICR (see Certification)

Electrical separation

This is a means of providing protection against shock from **one** item of equipment. (It can be used for more than one item, but the use of such an installation would need to be controlled or supervised by skilled persons, hence is quite rare.)

A typical example is a bathroom shaver unit, where the shaver is fed from the secondary side of an isolating transformer where there are no earths on the secondary side and hence it is electrically separate from the primary side.

Electricity at Work Regulations (EAWR) 1989

This is a **statutory** document that applies to all persons at work who are involved with electrical systems. Such systems, as defined by the EAWR, include anything from power stations to torch batteries, etc.

Contravention of certain of the Regulations may result in large fines, and in extreme cases imprisonment. Unlike all other legislation in the UK, persons who commit offences under the EAWR are presumed guilty and have to prove their innocence.

Electricity, Safety, Quality and Continuity Regulations (ESQCR) 2002

Formerly known as the Electricity Supply Regulations, the ESQCR are the province of the DNOs and require them to provide safe and standard supplies to consumers. They are also in a position to withdraw a supply to an installation if it is considered unsafe or could interfere with the public supply.

Electromagnetic compatibility (EMC)

Most modern electrical systems, and in particular electrical equipment, produce electromagnetic waves which, at a sufficiently high frequency, can cause malfunction of other equipment. We have all encountered the restriction in the use of mobile phones, for example, on aircraft. This electromagnetic interference (EMI) is becoming an increasing problem as more and more electronics impinge on our lives.

Hence there are a set of Regulations called 'The Electromagnetic Compatibility Regulations 2005', which provide requirements for electrical and electronic products in order to achieve electromagnetic compatibility.

Electromotive force (e.m.f.)

Measured in volts, this is the maximum voltage available in a cell/battery or generator to drive current around a circuit.

Once a load is connected and current (I) flows, the internal resistance (r) causes the voltage to drop to what is known as the 'terminal voltage' (V).

Hence a battery's terminal voltage **V** = battery e.m.f. **E** – internal volt drop **I × r**

$V = E - I \times r$

Electron

Electrons are the negatively (–ve) charged particles which form part of an atom, together with the associated positively (+ve) charged protons, and the neutrons, which have no charge. The flow of electrons in a circuit is known as current.

Enclosure (see also Barriers)

(BS 7671:2008 definition) 'A part providing protection against certain external influences and in any direction providing basic protection.'

Examples are consumer units, junction boxes, trunking, conduit, etc.

Equipotential bonding

In 1836 the physicist Michael Faraday had a large metal cage built, into which he, very sensibly, encouraged his assistant to enter !

The cage was then raised from the ground and charged to thousands of volts (gulp!).

His assistant found that he could move around the cage, simultaneously touching any parts, without any adverse effect. This was due to the fact that all parts were at the same potential and, as all 1st year electrical apprentices know (!), there needs to be a difference in potential for current to flow.

Bonding together all extraneous conductive parts of an installation with a main protective bonding conductor, which is connected to the main earthing terminal (MET), and having all exposed conductive parts also connected to the MET via cpcs, creates a Faraday cage in which we live or work.

Hence, in the event of a fault between line and earth, both exposed and extraneous conductive parts rise to the same potential.

There are some special situations when the risk of shock is greater which require supplementary equipotential bonding. Such locations include, for example, swimming pools, agricultural premises and circuses. Bathrooms are also included but such bonding may, under certain circumstances, be omitted.

Exhibitions, shows and stands

(BS 7671:2008 Section 711) 'This special location deals with display structures, etc. which are temporary in nature, and are typical of those found in, say, the Ideal Home or Electrex Exhibitions.'

Main points:

- Cables feeding temporary structures/stands/displays etc. must be protected at the supply end by 'time-delayed' or 'S' type RCDs not exceeding 300 mA, in order to give discrimination with other RCDs protecting final circuits
- Except for emergency lighting, all socket outlet circuits not exceeding 32 A must be protected by an RCD of maximum rating 30 mA
- Wiring cables shall be copper and of minimum csa 1.5 mm^2, and where there is a risk of mechanical damage, armoured cables must be used.
- The temporary electrical installation of structures/stands/displays etc. shall be inspected and tested ***on site after each assembly on site*** !

Exposed conductive part

(BS 7671:2008 definition) 'Conductive part of equipment, which can be touched and which is not normally live, but which can become live when basic insulation fails.'

Examples of such parts would be a metallic plate switch, the housing of a motor, the casing of a toaster, metal conduit and trunking, etc. In other words, Class I equipment and systems.

External influence (see also IP and IK codes, Barriers and Enclosures)

(BS 7671:2008 definition) 'Any influence external to an electrical installation which affects the design and safe operation of that installation.'

Such an influence could be, for example, the presence of water, dust, corrosion or impact, etc. In other words **environmental** conditions, or it could be those who use the installation or the materials that are used in it, i.e. the **utilization**.

Or the materials and structure of the **building**.

These three categories, **environment**, **utilization** and **building**, are allocated alpha-numeric codes to indicate the influence and its severity. Hence, environment is **A**, utilization is **B** and building is **C**. The second letter indicates the type of influence, and the number the severity.

So a code of AD4 indicates an environment subject to splashes whereas AD6 indicates waves. BD2 suggests difficult evacuation conditions, and CB3 a building subject to structural movement.

Appendix 5, BS 7671:2008 gives details of these external influences.

These codes are important, as they give an indication of the degree of protection that the equipment should provide against such influences in the form of the IP and IK codes.

Extra-low voltage, ELV (see also SELV, PELV, FELV and Band I)

(BS 7671:2008 definition) 'Not exceeding 50 V a.c. or 120 V ripple free d.c., whether between conductors or to earth.'

Extraneous conductive part

(BS 7671:2008 definition) 'A conductive part liable to introduce a potential, generally earth potential and **not** forming part of the electrical installation.'

Such parts would be, for example, structural steelwork, metallic gas, water and oil service and installation pipes. Basically, this is any metalwork that could be in contact with earth which would mean that it was at or near 0 volts. Such metalwork requires equipotential bonding.

Metal window frames (unless fixed into a building where the fixings are in contact with structural steel) are **not** extraneous conductive parts, nor are suspended ceiling grids, steel tables, metal chair legs, etc., i.e. metal parts that are not connected to earth.

A simple test to determine if an item is an extraneous conductive part is to measure the resistance between such a part and an earth point, say, for example the MET (main earthing terminal).

A value in excess of 22kΩ would suggest that the part was not extraneous.

F

Fairgrounds, amusement parks and circuses

(BS 7671:2008 Section 740) This section deals with temporary electrical installations for amusement devices and booths in fairgrounds, etc. It is not dissimilar to the requirements for exhibitions, shows and stands-it just contains more detail.

Main points:

- Automatic disconnection of supply must be provided at the origin by one or more 'time-delayed' or 'S' type RCDs not exceeding 300 mA in order to give discrimination with other RCDs protecting final circuits
- Additional protection by RCDs not exceeding 30 mA shall be provided for all final circuits for lighting, socket outlets not exceeding 32 A and flexible cables rated no more than 32 A feeding mobile equipment
- Additional protection by supplementary equipotential bonding shall be provided in areas intended for livestock
- All equipment must be to a minimum of IP44
- Socket outlets for use outside must be either BS EN 60903-1 or 2 except that socket outlets complying with a National Standard may be used, provided they give no less mechanical protection than a BS EN 60903-1 type and are 16 A maximum rating
- The electrical installation between the origin and equipment **must** be inspected and tested after each assembly on site.

Farad (symbol F)

This is the unit of capacitance and named after Michael Faraday, a British physicist and chemist (1792–1867), who is known as 'the father of electricity.'

Faraday cage (see Equipotential bonding)

Fault current I_f

(BS 7671:2008 definition) 'A current resulting from a fault.'

The Dictionary of Electrical Installation Work. DOI: B978-0-08-096937-4.00006-4

Such a fault could be either an earth fault or a short circuit fault.

Fault protection

(BS 7671:2008 definition) 'Protection against electric shock under single fault conditions.'

This protects against the risk of shock from contact with parts that have become live due to a fault (indirect contact) and is provided by:

1. Protective earthing, protective equipotential bonding and
2. Automatic disconnection in case of a fault.

FELV (functional extra-low voltage)

(BS 7671:2008 definition) 'An extra-low voltage system in which not all of the protective measures required for SELV or PELV have been applied.'

Such systems would include equipment that need ELV to **operate**, including transformers, relays, contactor coils, signal lamps, etc. Hence ELV is not necessarily used for safety purposes.

Note: When carrying out insulation resistance tests on FELV systems the requirements for LV systems must be met *(see Testing)*.

Ferromagnetic material

This is a material such as iron that is attracted to magnets and/or can become permanently magnetized.

Steel conduit and trunking is a ferromagnetic enclosure, and when there is a changing magnetic field close to it, small currents called 'eddy currents' are induced. These can combine to make quite large circulating currents which in turn produce heat.

If a single core line conductor, carrying a.c. current, is installed in the enclosure, there will be an alternating magnetic field which will induce eddy currents in the enclosure, causing it to heat up. If the circuits' associated neutral conductor is now included in the enclosure, the currents will flow in different directions as will their respective magnetic fields, which will cancel out and remove the eddy current effect.

BS 7671:2008 requires that, where conductors of an a.c. circuit are installed in a ferromagnetic enclosure, all line, neutral and protective conductors of that circuit should be combined in the same enclosure. Also when single core cables of an a.c. circuit enter a ferrous enclosure, for example a distribution board, the cables should all enter through the same entry hole. Sometimes this is not possible due to the hole size, and so slits are cut between adjacent holes. This prevents the build up of eddy currents in the enclosure around each cable.

First fix

This is the initial installation stage of erecting the wiring system prior to 'second fixing' of the accessories and/or equipment.

Flexible cables

These cables are constructed such that they may be flexed whilst in service.

Floor and ceiling heating systems

(BS 7671:2008 Section 753) The title here speaks for itself and applies to thermal storage or direct heating systems.

Main points:

- For automatic disconnection of supply, RCDs of maximum rating 30 mA shall be used as disconnecting devices
- Floor areas shall have a limiting temperature to prevent burns to skin (e.g. 30°C)
- To avoid overheating, the zones where heating units are installed should have a limiting temperature of 80°C
- Ceiling heating systems should be to at least IPX1 and floor heating to IPX7.

Fluorescent lighting

This comprises a luminaire, housing a ballast, starting arrangements, and a tube which contains mercury vapour at low pressure. The inside of the tube is coated with a fluorescent powder which converts the UV light produced to visible light *(see also Discharge lighting).*

Fly-lead (see Conduit and trunking)

FP cable

This is fire protection cable, the most common being FP200. It is used generally for fire protection and alarm systems.

Functional earthing

This is the earthing required to ensure that certain items of equipment work correctly. Such items tend to be related to information technology equipment. In these cases it is also likely that the earthing would be needed to provide protection for safety where there are high protective conductor currents.

Functional extra-low voltage (see FELV)

Functional switch

A switch that turns items or circuits 'on' or 'off' for operational purposes.

Fuses

Devices designed to operate when an overcurrent of sufficient value is reached, and which cause the melting of the fuse element within a specified time.

The most commonly used fuses are:

BS 88 gG......general purpose cartridge fuse
BS 88 gM......motor rated cartridge fuse
BS 1361........cartridge fuse for domestic and similar installations (typical service fuse)
BS 1362........fuse link for domestic and similar installations (typical plug fuse)
BS 3036........semi-enclosed rewireable fuse.

BS 88s and 1361s have a high breaking capacity and are often referred to as HBC fuses. They are necessary where the level of prospective fault current (I_{pf}) is high, generally close to the supply intake of an installation.

The breaking capacity of BS 88s can be as high as 80 kA and for BS 1361s, 30 kA.

The BS 3036 type has a poor value of up to only 4 kA.

Fuse switch

This is a switch that incorporates the fuse/s into the switch mechanism. Hence the fuse/s move with the switch blades *(see also Switch-fuse).*

Fusing factor

This is the ratio of the current rating of a protective device to its fusing or operating current.

For BS 88, BS 1361 fuses and circuit breakers, the fusing factor is generally accepted to be 1.45, in other words they can carry up to 1.45 times their rating before operating.

So a 10 A device can carry 14.5 A before it operates.

A BS 3036 rewirable fuse has a fusing factor of 2, so a 30 A device can carry up to 60 A before it operates.

A ratio of the two fusing factors, i.e. 1.45/2, results in a factor of 0.725 which is used in calculations to determine correct conductor sizes where BS 3036 fuses are used.

G

Gas installation pipe

These are the supply and installation pipes on the consumers' side of the gas meter.

Such a pipe is an extraneous conductive part and should, therefore, be bonded to the MET (main earthing terminal) with a main protective bonding conductor connected within 600 mm of the gas meter. If the meter is external, this conductor must pass through the wall in a separate hole to that of the gas pipe.

Note: Gas installation pipes must be at least:

1. 150 mm away from electricity meters, distribution boards, controls, switches and sockets, and
2. 25 mm from any cables.

General characteristics

This is the make-up of an installation and includes for example:

- The purpose of the installation
- Details of the DNOs supply, voltage, frequency, earthing, etc.
- External influences that may be relevant
- How the installation is to be divided into circuits
- The compatibility of the equipment and services, etc, etc.

Part 3 of BS 7671:2008 requires an assessment of such characteristics to be made before an installation is started.

GLS lamps

General Lighting Service lamps. These are the, soon to be obsolete, incandescent lamps we are all familiar with.

The Dictionary of Electrical Installation Work. DOI: 10.1016/B978-0-08-096937-4.00007-6

Ground/grounding

This is an American version of earth/earthing.

Grouping

Where multi-core cables or circuits are grouped together, they may impart heat to each other which may result in the need for a reduction in cable current carrying capacity.

BS 7671:2008 provides a table of rating factors to be applied to groups in order to determine the correct cable sizes.

Guidance notes GN (Institute of Engineering and Technology) (IET)

These are non-statutory documents published by the IET and provide guidance on the following:

GN 1...Selection and erection of equipment
GN 2...Isolation and switching
GN 3...Inspection and testing
GN 4...Protection against fire
GN 5...Protection against electric shock
GN 6...Protection against overcurrent
GN 7...Special locations
GN 8...Earthing and bonding

Guidance note (GS 38)

This is a non-statutory document published by the Health and Safety Executive (HSE) and is titled **"Electrical test equipment for use by electricians"**.

H

Harmonic currents

Items such as resistors, inductors and capacitors, which are linear loads, will produce a pure sinusoidal waveform known as the **fundamental**. Equipment such as switch mode power supplies (SMPS), electronic fluorescent ballasts, UPS (uninterupted power supply) and all other non-linear loads produce **harmonic** currents, in particular the 3rd harmonic. These add to the **fundamental** and cause a distortion. This distorted waveform may cause problems, such as circuit breaker tripping, overheating of transformers and neutral conductors; the latter being of particular concern in three phase systems where the neutral current can attain levels much higher that the line conductors.

This problem is becoming more pronounced as an increasing number of new electronic devices and control systems are introduced, and hence this topic requires careful consideration by designers of installations.

Harmonized document (HD)

This is a document that contains the technical details of a European Standard and is relevant to an individual country's national standard. For example HD 308 relates to BS EN 60446, the colour identification of conductors.

Health and Safety at Work Act 1974

This is a set of Statutory requirements intended to ensure safety in the workplace in all disciplines.

High protective conductor currents (see Protective conductor currents)

High voltage

(BS 7671:2008 definition) 'Normally exceeding low voltage. i.e. over 1000V a.c.'

Highway power supplies

These are the supplies to street located equipment, such as road signs, lamp-posts, etc.

The Dictionary of Electrical Installation Work. DOI: 10.1016/B978-0-08-096937-4.00008-8

HOFR cable

This cable is Heat and Oil resistant and Flame Retardant. It is generally used for the flexible cables used on construction sites, supply cable for mobile or transportable units, welding equipment, etc.

Hold-on circuit

This is a fundamental and important circuit arrangement. It is found in many applications ranging from alarm and security systems to motor control.

The diagram (Fig. 11) illustrates this simple circuit.

When the start button is pushed, the coil is energized and its normally open (N/O) contacts, 'A' and 'B', close. The load is then energized and the coil is 'held-on' via its own contact 'A' when the start button is released.

The load remains energized until the stop button is pushed, the coil is de-energized and the contacts return to their N/O positions. The load cannot re-start until the start button is operated. So simple but effective! *(see also Starter (motor).)*

Horsepower

One horsepower (hp) is equal to 746 watts of electrical power.

BS 7671:2008: 2008 requires that every motor that is rated in excess of 0.37kW, which is ½ hp, should have control equipment incorporating overload protection.

HSE (Health & Safety Executive)

This is the official organization whose inspectors police The Health and Safety at Work Act.

Hysteresis

A loss caused by an alternating magnetic flux in transformer and inductor cores is sometimes referred to as an 'iron loss'. Such losses, together with I^2 R losses, are referred to as 'copper losses', and they reduce efficiency.

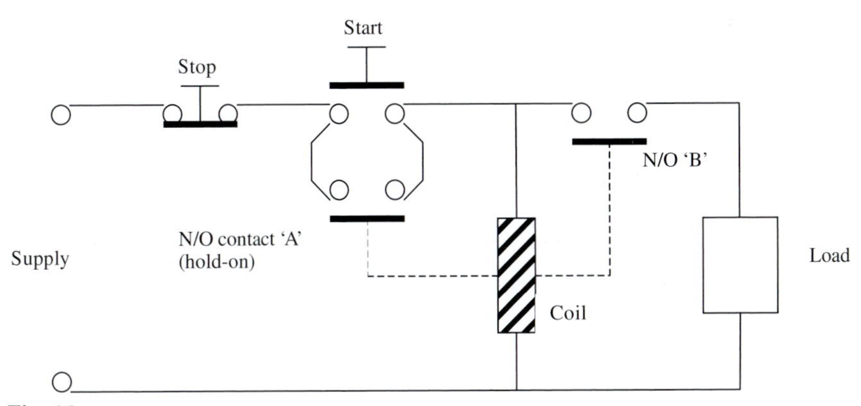

Fig. 11

I

I

I

This is the general symbol for current, in amperes.

I_a

This is the current that will cause automatic disconnection of a protective device within the time specified (e.g. 0.4 secs).

I_b (see Design current)

IEC (**International Electrotechnical Commission**) This is an organization that publishes and assesses conformity to standards for electrical, electronic and related technologies.

IET

The Institute of Engineering and Technology, formerly the IEE.

I_f (see Fault current)

IK codes (for impact)

These are a set of alpha-numeric codes denoting protection against severity of impact (Fig. 12).

BS 7671:2008 only mentions impact protection, using the reference IK08 or 5J (joule), in Part 7, special locations 705, 708, 709 and 740.

Impedance Z

This is an opposition to the flow of current, measured in ohms (Ω), and it is a combination of the effects of resistance, inductance and/or capacitance. It is given by:

$$Z = \sqrt{R^2 + X^2}$$

The Dictionary of Electrical Installation Work. DOI: 10.1016/B978-0-08-096937-4.00009-X

IK codes

Protection against mechanical impact

CODE		
00		No protection
01 to 05	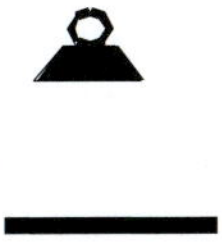	Impact < 1 joule
06	 500 g, 20 cm	Impact 1 joule
07	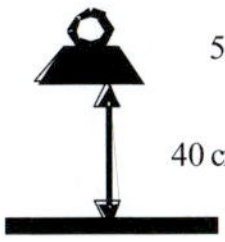 500 g, 40 cm	Impact 2 joule
08	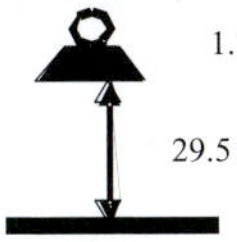 1.7 kg, 29.5 cm	Impact 5 joule
09	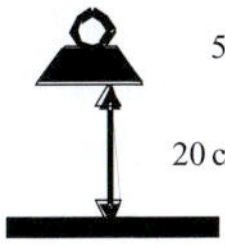 5 kg, 20 cm	Impact 10 joule
10	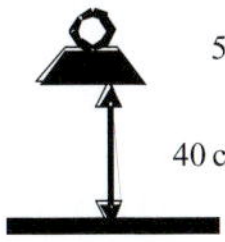 5 kg, 40 cm	Impact 20 joule

Fig. 12

Where R = Resistance in Ω

X = Either inductive or capacitive reactance X_L or X_C in Ω

or the difference between them, e.g. $(X_C - X_L)$ in Ω

Impulse withstand

When electrical installations are subjected to overvoltages due to lightning or switching surges, the equipment must be able to withstand such overvoltage impulses.

Equipment is categorized in BS7671 as category I, II, III or IV, and each category has a tabulated withstand voltage (kV).

Provided that the equipment product standard requires at least the tabulated value, then no surge protection would be needed.

(see also Surge protection)

I_n

The nominal current rating or setting of a protective device.

$I_{\Delta n}$

The residual operating current of an RCD.

Indirect contact (see Fault protection)

Induced emf

When an a.c. current flows in a conductor, an alternating magnetic field is produced around it. If the conductor is in close proximity to any ferrous material, e.g. iron or steel,the fluctuating magnet field causes an emf/voltage to be induced in that material. This may produce a potential shock risk, and, together with circulating currents known as eddy currents, it will produce heat.

If, however, another conductor is run with the first, and it carries current in the opposite direction, the magnetic effects are cancelled out and no induced emfs/voltages/eddy currents will occur.

Hence, the line and neutral conductors of a circuit should be run together in the same metal conduit or trunking. They carry current in opposite directions! Hence there can be a problem with what is known as 'borrowed neutrals' where a neutral is taken, for convenience reasons, from a different circuit to that of the line conductor.

(see also Ferromagnetic material)

Inductance L henries

This is the property of an a.c. circuit that produces (magnetic) opposition to the flow of current in that circuit.

Inductive reactance X_L ohms

This is an opposition (magnetic) to current flow in an a.c. circuit. *(See also Back emf)*

Inductor

This is an iron cored coil, in electrical installations more commonly known as a choke or ballast, found in discharge luiminaires, e.g. fluorescent fittings.

Initial verification (see also Testing)

This is the inspection and testing of a new installation, or an alteration or addition to an existing installation.

Inspection

This is the part of the verification process which should be carried out before testing, and normally with that part of the installation that is under inspection being disconnected from the supply.

Instructed person

(BS 7671:2008 definition) A person adequately advised or supervised by skilled persons to enable him/her to avoid dangers that electricity may create. Such persons could include, for example, mechanical maintenance staff, school caretakers, apprentices, etc.

Insulation monitoring device (IMD)

These devices are used primarily in IT earthing systems, where disconnection of the supply is not permitted, and hence are not generally used in everyday installations. Medical locations, particularly operating theatres, etc., are typical examples of where they may be needed. IMDs continually monitor the condition of insulation resistance, and provide an audible and visual indication if the resistance falls below a pre-set level.

Insulation resistance (see Testing)

Interconnection diagram (see Diagrams)

Intumescent materials

These are materials that expand in volume when heated; hence they are well suited to filling holes where cables have passed through fire resistant walls and ceilings. Recessed downlighters often require intumescent hoods.

IP codes

These are codes that indicate how well a barrier or enclosure can protect against water and/or foreign solid bodies.

▼ Index of Protection (IP) Code

First numeral		Second numeral	
(a) Protection of persons against contact with live or moving parts inside enclosure (b) Protection of equipment against ingress of solid bodies		Protection of equipment against ingress of liquid	
No./Symbol	Degree of protection	No./Symbol	Degree of protection
0	(a) No protection (b) No protection	**0**	No protection.

continued on the next page

First numeral		Second numeral	
1	(a) Protection against accidental or inadvertent contact by a large surface of the body, e.g. hand, but not against deliberate access (b) Protection against ingress of large solid objects <50 mm diameter	1	Protection against drops of water. Drops of water falling on enclosure shall have no harmful effect
2	(a) Protection against contact by standard finger (b) Protection against ingress of medium size bodies <12 mm diameter, 80mm length	2	Drip Proof Protection against drops of liquid. Drops of falling liquid shall have no harmful effect when the enclosure is tilted at any angle up to 15° from the vertical
3	(a) Protection against contact by tools, wires or suchlike more than 2.5 mm thick (b) Protection against ingress of small solid bodies	3	Rain Proof Water falling as rain at any angle up to 60° from vertical shall have no harmful effect

continued on the next page

First numeral		Second numeral	
4	(a) As 3 above but against contact by tools, wires or the like more than 1.0mm thick (b) Protection against ingress of small foreign bodies	**4**	Splash Proof Liquid splashed from any direction shall have no harmful effect
5	(a) Complete protection against contact (b) Dustproof Protection against harmful deposits of dust. Dust may enter but not in amounts sufficient to interfere with satisfactory operation	**5**	Jet Proof Water projected by a nozzle from any direction (under stated conditions) shall have no harmful effect
6	(a) Complete protection against contact (b) Dust-tight Protection against ingress of dust	**6**	Watertight equipment Protection against conditions on ships' decks, etc. Water from heavy seas or power jets shall not enter the enclosures under prescribed conditions

continued on the next page

First numeral	Second numeral	
IP CODE NOTES - Degree of protection is stated in form IPXX - Protection against contact or ingress of water respectively is specified by replacing first or second X by digit number tabled, e.g. IP2X defines an enclosure giving protection against finger contact but without any specific protection against ingress of water or liquid Codes may also be explained with letters e.g. IP2X or IPXXB IP4X or IPXXD	**7**	Protection against immersion in water- It shall not be possible for water to enter the enclosure under stated conditions of pressure and time
	8	Protection against indefinite immersion in water under specified pressure It shall not be possible for water to enter the enclosure

If, for example, an item of equipment is to be used in an environment subject to jets of water and dust, it would be coded IP 55.

Isolation

This is the cutting off, for reasons of safety, all or part of an electrical installation from all sources of supply.

Isolator

A mechanical switching device used for the purpose of isolation, also known as a **disconnector**.

I_t

The minimum tabulated value of current carrying capacity of a conductor derived by dividing I_n or, where a circuit does not need overload protection, I_b, by **relevant** rating factors:

$$I_t \geq \frac{I_n \text{ or } I_b}{Ca \times Cg \times Cf \times Cc \times Cs \times Ci \times Cd}$$

(See Rating factors)

IT system (see Earthing systems)

J

Joint Industry Board (JIB)

This is the electrical contractors' national organization that sets wage rates, qualifications and grading, etc., for its members. It basically provides a liaison between unions and employers. Hence, a JIB graded electrician would have attained a specific qualification level and would be paid accordingly.

Joule symbol J

This is the unit of work and energy named after James Prescott Joule, a British scientist/engineer (1818–1889).

The Dictionary of Electrical Installation Work. DOI: 10.1016/B978-0-08-096937-4.00010-6

K

k values for conductors

A tabulated factor based on the materials from which a cable (insulation and conductor) is constructed and used together with the conductor cross sectional area '**S**' to give the heat energy that a cable can withstand, $\mathbf{k^2S^2}$.

(see also the Adiabatic equation and Let-through energy)

The Dictionary of Electrical Installation Work. DOI: 10.1016/B978-0-08-096937-4.00011-8

L

Leakage current

(BS 7671:2008 definition) 'Electric current in an unwanted conductive part under normal operating conditions.'

This could be current flowing across insulation, or through the screening of co-axial cables, or in protective conductors associated with IT equipments, etc. In the latter case it is referred to as 'protective conductor current'.

Let-through energy

When a fault current (I) occurs, a fuse element ruptures or a circuit breaker contact parts, and an arc is drawn between the broken or parted ends. This arc is extinguished, dependent on the protective device construction, over a short period of time (t). During this time electrical energy is 'let-through' the protective device into the cable. This energy is determined by the formula $\mathbf{I^2t}$.

Line conductor

This is a conductor for transmitting electrical energy other than a neutral conductor. In other words it's the brown, black or grey one we associate with single and three phase circuits.

Live conductor/part

(BS 7671:2008 definition) 'A conductor or conductive part intended to be energized in normal use including a **neutral** conductor but, by convention, not a PEN conductor.'

Loop impedance (see Earth fault loop impedance)

Low voltage, LV (see also Band II)

(BS 7671:2008 definition) 'Exceeding extra-low but not exceeding 1000V a.c. or 1500V d.c. between conductors, or 600V a.c. or 900V d.c. between conductors and earth.'

The Dictionary of Electrical Installation Work. DOI: 10.1016/B978-0-08-096937-4.00012-X

Unfortunately there is much confusion between LV and ELV, which is not helped by many manufacturers, especially those concerned with lighting, who insist on referring to their 12V systems as 'Low voltage lighting.' This gives purchasers/consumers the false idea that a low voltage is safe!

LSF cable

This is 'low smoke and fume' cable, generally used in public areas such as theatres, museums, etc. In the event of a fire the cable will not produce heavy fumes and black smoke.

Luminaire

Basically this comprises all the component parts that make up a light fitting, with the exception of the lamp or lamps themselves.

Lumen lm

This is the measure of the amount of light that flows from a light source.

Lux lx or lumens/m^2

No! Not soap powder! This is the unit for the amount of light that falls on a surface and is sometimes referred to as 'illumination.'

M

Magnetic field

This is the concentration of lines of force that surround a magnetic source. It is important to ensure that such fields do not interfere with the correct operation of equipment, and in particular RCDs.

(see also Electromagnetic influences)

Main earthing terminal (MET)

This is the earth bar in the supply distribution board and, where required for ease of disconnection for test purposes, an external block for the connection of protective bonding conductors, etc.

The MET is provided for the connection of:
Circuit protective conductors
Protective bonding conductors
Function earthing conductors (if required)
Lightning protection system bonding conductor (if any).

Marinas

(BS 7671:2008 Section 709) This section deals with circuits which are intended to supply pleasure craft and houseboats, with the exception of houseboats that are directly supplied from the public network.

Main points:

- Electrical equipment should withstand the external influences of water, foreign solid bodies and impact by ensuring it is coded at least IPX4, IPX5, IPX6, IP3X and IK08 respectively, and protected against corrosion where extevnal influences AF2 or AF3 are present
- Overhead cables should be 6 m above ground level in vehicle movement areas and 3.5 m in all others, and support poles should be placed to avoid damage

The Dictionary of Electrical Installation Work. DOI: 10.1016/B978-0-08-096937-4.00013-1

- Underground cables, unless provided with additional mechanical protection, should be at a suitable depth to prevent damage; 500 mm is considered to be a minimum
- Socket outlets should be:
  1. To BS EN 60309-2, up to 63A, BS EN 60309-1, above 63A and IP44 rated
  2. One per pleasure craft or houseboat
  3. Individually protected against overcurrent
  4. Individually protected by a 30mA or less RCD
  5. Not less than 1m above the highest water level. This may be reduced to 300 m for floating pontoons or walkways only, provided that the effects of splashing are taken into account.

Maximum demand (see also Diversity)

This is the sum of all the full loads of equipment and circuits in an installation.

Medical locations

(BS 7671:2008 Section 710) These locations include hospitals, private clinics, dental practices, healthcare centres, etc. Installations in such establishments are inevitably complex and hence only a brief overview is considered here.

Rooms in medical locations are divided into groups 0, 1, and 2. Group 0 are massage rooms, group 1 are general treatment rooms, such as physiotherapy, hydrotherapy, radiology, etc. Group 2 are operating and intensive care rooms.

Main points:

- Safety sources feeding essential services must be provided – i.e. standby systems
- Where required, in groups 1 and 2 only, type A or B RCDs are permitted
- For TN systems, final circuits up to 63A shall have protection by 30mA or less RCDs
- IT systems shall be used for equipment and systems in group 2 locations which are intended for life support and surgical applications
- Supplementary equipotential bonding shall be provided in each medical location of groups 1 and 2
- Unwanted tripping of 30 mA or less RCDs must be taken into consideration
- For IT systems in group 2, socket outlets supplying medical equipment must be un-switched and fitted with a supply indicator
- For initial verification, the additional functional tests of 'insulation monitoring devices' and 'overload monitoring' together with the verification of the integrity of the supplementary bonding must be carried out.

Miniature circuit breakers (see Circuit breakers)

Minor Electrical Works Certificate (see Certification)

Minor works

(BS 7671:2008 definition) 'Additions and alterations to an installation that does not extend to a new circuit.,

Mobile and transportable units

(BS 7671:2008 Section 717) These include units which are used for fire fighting, medical services, catering, etc., all of which have internal wiring for current-using

equipment, socket outlets, etc. They may have their own generators, or require a supply from an external source.

Main points:

- Automatic disconnection of supply shall be provided by an RCD
- Unless the installation is under the supervision of a skilled or instructed person, and the earthing has been confirmed to be suitable and effective, an external supply shall **not** be taken from a PME system
- Socket outlets used for supplying equipment outside the unit must be protected by an RCD of maximum rating 30mA and be IP44 rated
- Flexible supply cables (for connecting the unit to the supply) shall be a minimum of $2.5mm^2$ copper.

Mobile equipment

(BS 7671:2008 definition) 'Electrical equipment which is moved while in operation or which can easily be moved from one place to another while connected to the supply.' Apparently the IET disapprove of the term portable equipment, with which everybody else is familiar!

Modular wiring system

This is a pre-fabricated system used in commercial and industrial installations. It comprises distribution boards, special connection units, outlet boxes, and pre-set flexible cable lengths that have plugs on each end (Fig.13).

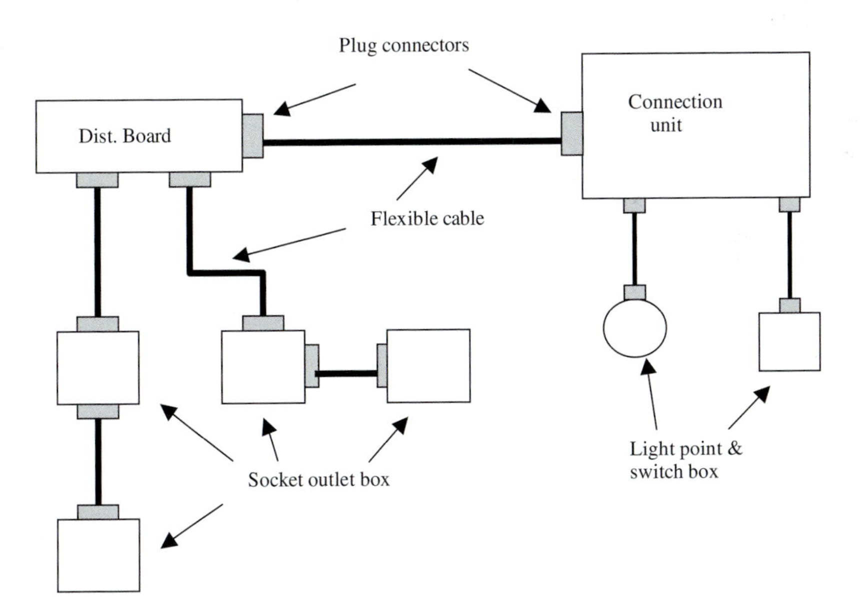

Fig. 13

It requires little effort to erect; the installer just fixes the various connection and outlet units and plugs the cable in at both ends. This clearly avoids the need to cut cable from a reel and terminate the ends. However it clearly needs good designing to ensure the cables are manufactured to the correct lengths.

Mutual detrimental influence

This refers to a situation where different electrical systems or electrical and non-electrical equipment may have harmful effects on one another, e.g. gas and electricity services *(see also Segregation).*

N

NAPIT

This is the National Association of Professional Inspectors and Testers whose aim is to ensure that its members are competent to provide a quality service to consumers. It is similar to the NICEIC. It is an approval body for Part 'P' of the Building Regulations.

Neutral conductor (see also Harmonics)

Such an understated conductor! As a **live** conductor it carries current in normal service. It is the return conductor in single and three phase installations, and in the latter it carries **out-of-balance** currents.

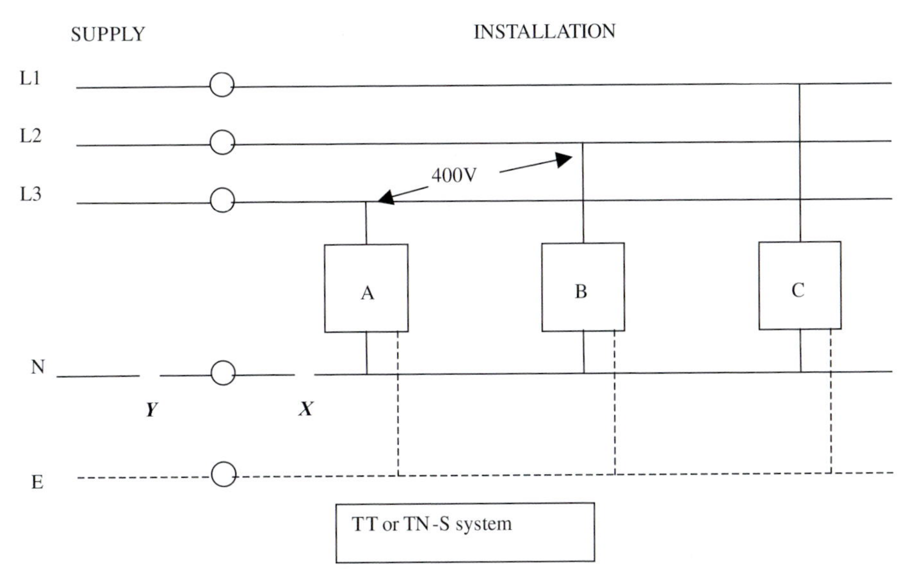

Fig. 14

The Dictionary of Electrical Installation Work. DOI: 10.1016/B978-0-08-096937-4.00014-3

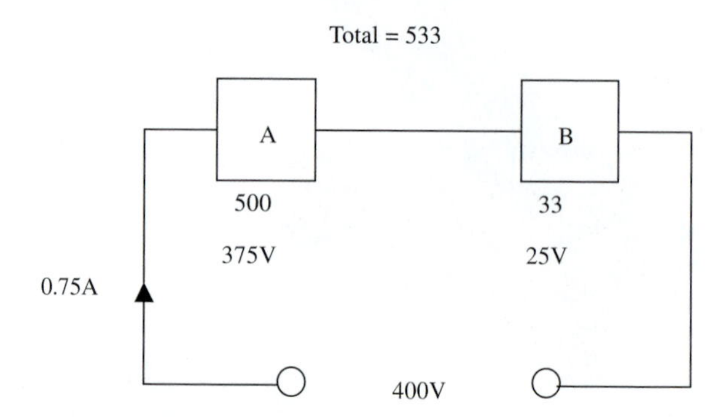

Fig. 15

Without it, single phase loads will just not work, whereas in three phase installations its loss can have serious consequences. It is even more serious if a neutral is lost on the supply side of a TN-C-S system. The following examples/diagrams may help to illustrate the problems of lost neutrals.

A break in the neutral at X or Y will result in the 230V loads (A and B) or (B and C) or (A and C) being connected in series across 400V (Fig. 14).

Using A and B as an example (Fig. 15), if load A was a computer rated at, say, 106W, it would have a resistance of approx. 500 Ω, and if B were a 16kW heater, it would have a resistance of approx. 33 Ω.

This means that the total current would be 400/533 = approx 0.75A.

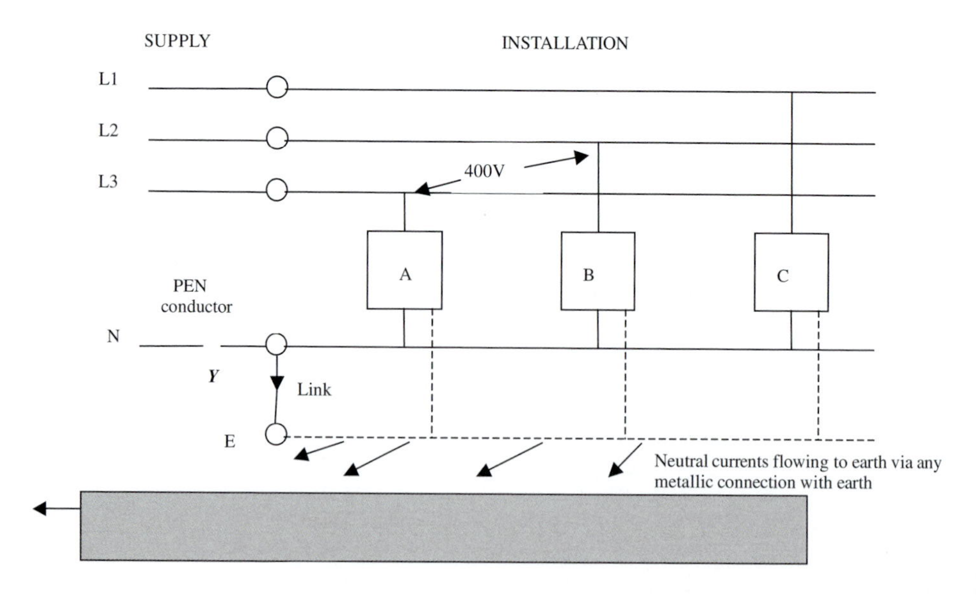

Fig. 16

So the voltage across the computer would be 0.75 × 500 = 375V; not the best situation for a 230V electronic item!

This situation is not just restricted to three phase installations; it can occur in houses or flats which are supplied from different phases of the Distribution Network Operators three phase system.

The consequences can be much worse on a TN-C-S system when the neutral is lost on the supply side of installations (Fig. 16).

With a break at Y, the installations' earthing will become live via the link, as the load currents attempt to flow back to the transformer through any earth paths they can find. This can be dangerous where there are metal, gas and oil service pipes, as there is a risk of fire or explosion. Note: A, B and C could just as easily be individual dwellings! *(See also Overvoltage)*

Another situation which can cause problems is 'borrowing a neutral' *(see Induced EMF).*

Neutrons

These are sub-atomic particles that have no electric charge which, together with positively (+ve) charged protons, form the nucleus of an atom. They play no part in general electrical engineering.

Newton symbol N

This is the unit of force or effort, named after Sir Isaac Newton, British physicist (1643–1727). A load force of 1N is equivalent to a mass of 9.81kg

NICEIC

This is the National Inspection Council for Electrical Installation Contractors. It concerns itself with maintaining the standards of its members' workmanship. It is an approval body for Part 'P' of the Building Regulations.

Non-conducting location

This, as the name suggests, is a location where floors and walls, etc are constructed of non-conducting materials. Such locations are uncommon and include areas such as specialist medical treatment rooms, or storage areas for explosives.

Notices

The IET Wiring Regulation requires that notices be displayed where relevant throughout an installation *(see also Warning notices).*

Nuisance tripping

This is the term commonly used for the unwanted operating of a protective device. Such operation is generally associated with RCDs which are so very sensitive that they can trip due to the starting characteristics of items of equipment like fridges/freezers. In fact anything that causes a momentary out-of-balance in the RCD will cause this irritating situation.

Circuit breakers (cbs) are also prone to nuisance tripping, although less so than RCDs. Cbs may sometimes operate when a tungsten filament lamp fails, as this causes a momentary overload.

If a circuit breaker operates when the motor circuit it is protecting is energized, it is probably the type of breaker that is incorrect rather than nuisance tripping (although it may be a bit of a nuisance!)

O

Obstacles

An obstacle in an installation may be used a means of basic protection, in that it is intended to prevent unintentional contact with live parts. An example of an obstacle would be a handrail located in front of an open fronted bus-bar chamber in a switch-room. Because of the obvious dangers associated with this arrangement, this method of protection can only be applied where the installation is controlled or supervised by skilled persons.

(see also Operating and maintenance gangways)

Ohm symbol Ω

This is the unit of electrical resistance, and it is named after the German physicist Georg Simon Ohm (1789–1854), who demonstrated the relationship between current, voltage and resistance.

Ohm's Law

This is the basis on which our understanding of electricity is built. It states:

'The current in a circuit is proportional to the circuit voltage and inversely proportional to the resistance, at constant temperature'

This gives rise to the familiar formulae **$I = V / R$ or $V = I \times R$ or $R = V / I$**, which are often expressed for ease using the triangle shown in Fig. 17.

Operating or maintenance gangways

(BS 7671:2008 Section 729 and definition) Gangway providing access to facilitate operations such as switching, controlling, setting, observation and maintenance of electrical equipment.

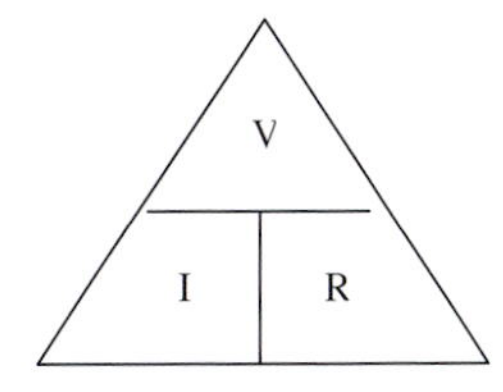

Fig. 17

The Dictionary of Electrical Installation Work. DOI: 10.1016/B978-0-08-096937-4.00015-5

Such gangways are likely to be found in restricted areas which are typical of switchrooms, etc. where protection against contact with live electrical parts of equipment is provided by barriers or enclosures or obstacles, the latter having to be under the control of skilled persons.

Main points:

- Restricted areas must be clearly and visibly marked
- Access to unauthorized persons is not permitted
- For closed restricted areas, doors must allow for easy evacuation without the use of a key or tool
- Gangways must be wide enough for easy access for working and for emergency evacuation
- Gangways must permit at least a 90° opening of equipment doors etc.
- Gangways longer than 10 m must be accessible from both ends.

Ordinary person

(BS 7671:2008 definition) 'A person who is neither a skilled person nor an instructed person.'

In this context, skilled or instructed refers to matters electrical.

Overcurrent

(BS 7671:2008 definition) 'A current exceeding the rated value. For conductors, the rated value is the current-carrying capacity.'

Overcurrent is a generic term and includes overload, earth fault and short-circuit currents.

Overload current

(BS 7671:2008 definition) 'An overcurrent occurring in a circuit which is electrically sound.'

So, the wiring is all ok, it's just that equipment is causing excess current. This could be due to a faulty item such a motor stalling or too many appliances being connected to a circuit.

Overload protection

This is generally provided by fuses, circuit breakers or thermal devices such as bi-metal overloads in motor starters.

Overvoltage

A temporary overvoltage in an installation may be caused by:

- Earth faults on the high voltage system
- Short circuits on the low voltage installation
- Loss of the neutral in the low voltage system.

A transient (short lived) overvoltage in an installation may be caused by:

- Atmospheric origin (lightning)
- Switching surges generated by equipment in the installation.

P

Parallel conductors/cables

There are occasions when two or more conductors/cables are wired in parallel. This is usually done where the load current is larger than is either practical or economical for just one to be enough. In these circumstances the conductors/cables should ideally be of the same material and cross sectional area so that the load is shared equally.

There are occasions when additional protective conductors are wired in parallel in order to reduce R2 and hence Zs values.

Another common use of parallel cables is the ring final circuit.

Parallel paths

These can have both adverse and beneficial effects in installation work.

When testing for protective conductor resistance, parallel paths caused by other circuits or extraneous conductive parts such as pipework, may give optimistically low values. This may mislead those testing into believing that the protective conductor under test is continuous and soundly connected. This is of particular significance with protective bonding conductors which should be disconnected before testing.

Care must be taken to ensure that the supply is isolated before such disconnection and that the conductor is reconnected on completion of the test.

Conversely, these parallel paths are of added benefit under earth fault conditions as Zs values will be lower than design values ensuring faster protection disconnection times than calculated.

The Dictionary of Electrical Installation Work. DOI: 10.1016/B978-0-08-096937-4.00016-7

Part 'P'

This is the part of the Building Regulations that deals with the safety of electrical installations in dwellings. It requires that such installations be designed and installed in such a way as to protect persons from fire and injury; in other words they must conform to BS 7671:2008.

Approved Document P provides guidance on how this requirement may be achieved.

One such item of guidance suggests that the design and installation be carried out by a competent person belonging to an approved Competent Persons Scheme and who is registered as a Domestic Installer.

Peak value (see a.c.)

PEN conductor

(BS 7671:2008 definition) 'A conductor combining the functions of both protective conductor and neutral conductor.'

This is the 'protective earthed neutral' conductor provided by the Distribution Network Operator and forms part of a TN-C-S system, i.e. the PME part.

PEN conductors are not permitted in a consumer installation unless specific authorization is obtained from the ESQCR, or the installation is fed from a private source.

PELV (protective extra-low voltage)

PELV is used as some measure of protection against electric shock. It fulfils the requirements for both **basic** and f**ault** protection. It does not have the same degree of safety as SELV as there may be connections to earth on the secondary (ELV) side of the safety isolating transformer.

It is however generally accepted as an alternative to SELV as will be seen throughout BS 7671:2008 where many Regulations require the use of either SELV or PELV.

An example of the use of PELV would be for a fire alarm system where cables and call points, etc. may have connections to earth.

(see also SELV)

Periodic inspecting and testing

This is carried out in order to ensure that an installation or part of it is in a satisfactory condition for continued use *(see also Electrical Installation Condition Report, under Certification).*

Phase

Used to denote equipment or systems, e.g. single **phase** motor or three **phase** supply.

The use of the word **phase** when referring to conductors has been replaced by **line.**

For example a three **phase** three wire system would have three **line** conductors

Phase sequence (see Testing)

Placing out of reach

This is a method of providing basic protection. Live parts are located from a work surface:

- 2.5 m standing
- 1.25 m lying (reaching below).

Such a method is uncommon and can only be used when the installation is controlled or supervised by a skilled person *(see also Arm's reach).*

Point (in wiring)

(BS 7671:2008 definition) 'A termination of the fixed wiring intended for the connection of current-using equipment.'

Typical examples include socket outlets, ceiling roses/lampholders, isolators, connection units, etc.

Polarity (see Testing)

Portable equipment (see Mobile equipment)

Potential difference (p.d.) (volts)

This is the difference in voltage between two points. Hence in the UK, 230 V is the p.d. between line and earth (U_0).

Power P (watts)

P

This is the ability of electrical systems to convert potential energy into work, and is the product of the current I and the voltage V, hence:

$$P = I \times V$$

As Ohm's law gives us $V = I \times R$ or $I = V / R$, so we can also express P as:

$$P = I^2 R \text{ or } P = V^2 / R$$

In single-phase circuits the power is given by:

$$P = I_b \times V \times p.f \times Eff\,\% \text{ watts}$$

Where I_b = design current in amperes

V = voltage (usually 230 V)

p.f. = power factor

In three-phase circuits the power is given by:

$$P = \sqrt{3} \times I b \times V L \times \text{p.f.} \times \text{Eff}\ \%\ \text{watts}$$

Where $\sqrt{3} = 1.732$

I_b = design current in amperes

V_L = line voltage (usually 400 V)

p.f. = power factor

Eff % = efficiency expressed as a percentage.

Power factor (p.f.)

This is probably best explained by using the 'beer analogy'. Consider a pint glass of beer. In some pubs there will be a lot of froth or head; in others, very little. The pint you pay for, the 'apparent pint', is made up of the actual beer, the 'true pint' and the froth or 'wasted pint'.

If you now do a ratio of *true* pint to *apparent* pint you will have a ***pint factor***, hence:

$$\text{Pint factor (pf)} = \frac{\text{true pint}}{\text{apparent pint}}$$

A pint factor (p.f.) of 1 means no froth, whilst a p.f. of 0.5 means half froth and half beer – a pub to avoid!

A p.f. of 0.95 is much more sensible although somewhat hopeful! So a 'pint factor' indicates the actual beer you get and what you are really paying for.

The beer analogy gives an **idea** of the electrical power factor, which is similar but slightly more complicated. Where electrical equipment is concerned there are, similarly, three types of power: true, apparent and wasted

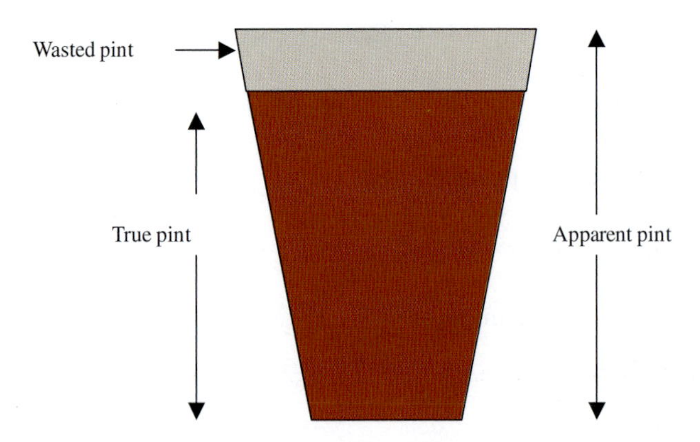

Fig. 18

1. True power, usually measured in kilowatts (kW), is associated with resistive items such as cables and heaters.
2. Apparent power, usually measured in kilovolt amperes (kVA), is associated with electromagnetic or reactive items such as transformers, motors, ballasts, etc.
3. Wasted power, usually measured in kilovolt amperes reactive (kVAr), is associated with magnetism and is a bit like the beer froth.

Electrical power factor is just like the pint factor, in that it is also the ratio of 'true' to 'apparent' hence:

$$\text{Power factor (p.f.)} = \frac{\text{true power}}{\text{apparent power}} = \frac{\text{kW}}{\text{kVA}}$$

In the same way as the beer, kW is what you are usefully using but kVA is what you pay for.

However, unlike the apparent pint, kVAr (wasted power) cannot simply be added to the kW (true power) to obtain the kVA (apparent power). These values are linked by the formula:

$$\text{kVA} = \sqrt{\text{kW}^2 + \text{kVAr}^2}$$

An example of power factor in everyday use is that of the fluorescent luminaire – or in fact any discharge lighting unit. The ballast or choke in the unit presents a reactive component to the power and hence, unless the actual p.f. of the unit is known, the VA rating is obtained by multiplying the lamp watts (W) by 1.8.

So a 1200 W fluorescent luminaire would have a rating of $1200 \times 1.8 = 2160$ VA. Hence the current rating would be calculated from:

$$I_b = \text{VA} / \text{V} = 2160 / 230 = 9.4\,\text{A}$$

not $\text{W} / \text{V} = 1200 / 230 = 5.2\,\text{A}$

(see also Discharge lighting)

Power factor correction

This is nearly always achieved by the use of capacitors connected across the supply terminals of equipment, circuits or installations. Although it is theoretically possible to correct a power factor to unity, i.e. 1, the cost of capacitors needed to achieve this would be prohibitive, as any saving in energy costs would be lost. In general p.f. is corrected to around 0.9.

In some very large installations with a considerable number of inductive loads, banks of power factor correction capacitors situated at the origin of the installation automatically switch on and off with variation in the loads.

PPE

The Personal Protective Equipment at Work Regulations 1992.

Prospective fault current (I_{pf})

Imagine, say, a socket outlet at the end of a radial circuit. Then imagine that all the protective devices right back to the supply transformer were replaced with solid links. Then imagine that a dead short was made at the socket between either the line and neutral, or the line and earth. The current that would flow is the **prospective fault current**. The first, L to N, is a 'short circuit current', the second, L to E, is an 'earth fault current'.

The closer we get to the origin of an installation the higher is the fault current, and the farther away, the lower is the fault current. (Remember that an increase in resistance causes a decrease in current!)

Protective devices must have a breaking capacity suitable for I_{pf} at the point that it is installed *(see also Breaking capacity, Circuit breakers and Fuses)*.

This requires employers to provide appropriate PPE for their employees.

Protection

See fuses, circuit breakers, IP and IK codes, equipotential bonding.

Protective conductors

These conductors are used for some measure of protection against electric shock and include:

1. The earthing conductor
2. Main protective bonding conductors
3. Supplementary protective bonding conductors
4. Circuit protective conductors (cpcs)

(see also Equipotential bonding)

Protective conductor current

(IET Regulation definitions) Electric current appearing in a protective conductor such as leakage current or electric current resulting from an insulation fault.

This is typical of leakage currents caused by IT equipment, computers, data processing, etc.

Where there are large quantities of such equipment, for example computers in a large office, the combined protective conductor current can reach significant levels, and BS 7671:2008 requires specific measures regarding earthing to be carried out.

Also, high values of such current may cause unwanted tripping of RCDs *(See Nuisance tripping)*.

Protective extra-low voltage (see PELV)

Protective multiple earthing (PME)

This is the Distribution Network Operators (DNOs) part of a TN-C-S system *(see also Earthing systems and Neutral conductor)*.

The neutral conductor of a DNOs distribution cable is connected to earth at many points along its length in order to get the potential as close to 0 volts as possible.

It must be remembered that this 'earthed neutral' provides an **artificial** earth to the consumer. It should not, ideally, be extended outside an installation as it may, in the event of a line/neutral fault on the DNOs system, present a voltage to the user of Class 1 equipment, which could be dangerously above true earth potential.

In fact the ESQCR prohibit the use of TN-C-S systems in certain circumstances such as construction sites, caravan parks, marinas, etc.

Protons

These are positively (+ve) charged particles which, together with neutrons, make up the nucleus of an atom.

PUWER

The Provision and Use of Work Equipment Regulations 1998.

These ensure the safety of work equipment on site.

PV supply systems

(BS 7671:2008 Section 712) These comprise photovoltaic (PV) solar panels and d.c. to a.c. invertors together with, where relevant, transformers, isolators, protective devices and associated wiring. Such systems provide a renewable energy source and are increasingly used to supplement standard supply sources.

R

R1 + R2

This is the sum of the resistances of a circuit line conductor (R1) and its associated cpc (R2).

Values for R1 and R2 may be found in the IET's 'On-site-guide' or Guidance Note 1. They are quoted in milliohms (mΩ) per metre (m) at 20°C and will therefore need correcting for the operating temperature of the cable in question.

Multipliers are given for this correction; 1.2 for 70°C thermoplastic insulation, 1.26 for 90 °C thermoplastic and 1.28 for 90°C thermosetting.

(see also Temperature coefficient and Earth fault loop impedance)

Radial circuits (see Circuits)

Radio interference suppression

R

Rapidly and continually switching a circuit on and off can cause constant arcing across the switch contacts. Such switching causes frequencies that can interfere with radio and television reception. A typical example of this occurs in a fluorescent luminaire starter. A small capacitor is built into the starter across the starter contacts which cuts out the interference.

Ramp test (RCD)

This test is available on some RCD test instruments. The instrument performs a series of automatic incremental tests between $1/2I\Delta_n$ and $1\ I\Delta_n$, and indicates the actual current that trips the RCD. It is a useful indication of the susceptibility of the RCD to nuisance tripping.

Rating factor C

Rating factors are applied during the calculation of cable current carrying capacity I_t.

The Dictionary of Electrical Installation Work. DOI: 10.1016/B978-0-08-096937-4.00017-9

Their application accounts for the various conditions to which a cable is subjected along its route. Such conditions are 'ambient temperature', 'grouping or bunching of cables', use of BS 3036 rewirable 'fuses', 'burying in ducts or directly in the ground', 'depth of burial', 'soil resistivity' and 'contact with thermal insulation'.

They are classified as follows:
C_a .....Ambient temperature (found from BS7671 Tables 4B1)
C_g .....Grouping (found from BS7671 Tables 4C1)
C_f .....BS 3036 fuse.. **0.725**
C_c .....Buried cable.. **0.9**
C_d .....Depth of burial
C_s .....Soil resistivity
C_i .....If totally surrounded with thermal insulation for more than 0.5m...**0.5**.
If less than 0.5, found from BS7671 Table 52.2.

The relevant factors are applied as divisors to the rating of the protective device I_n or design current I_b if the circuit is not subject to overloads *(see also I)*.

RCBO

This is an RCD with integral overload and short circuit protection, i.e. a combined RCD/cb.

Some RCBOs have a white fly lead. These are the electronic types (generally type A) and the white lead is connected to the neutral bar. It does not have to be connected, but without it earth leakage will not be detected if there is a loss of the supply neutral.

RCCB

This is an RCD without integral overload and short circuit protection.

RCD (see Residual current device)

This is a generic term and includes RCBOs and RCCBs.

RCD testing (see Testing)

Reduced low voltage system

This system is generally used on, but not exclusive to, construction sites. A centre tapped to earth (CTE) step-down transformer reduces the voltage from 230 V to 110 V. In the case of single-phase, the voltage on each side of the centre point is 55 V to earth. With three-phase, the voltage between the mid or star point and each phase is 63.5 V.

This system comes under the measure of protection of 'Automatic Disconnection of Supply' and, as with 230V systems, there are maximum values of earth fault loop impedance. The disconnection time for these values is 5 seconds.

Reinforced insulation

This is insulation applied singly or in layers to live parts and designed to provide the same degree of protection as double insulation.

(see also Class II equipment and Supplementary insulation)

Residual current device (RCD) (see also Additional protection and Nuisance tripping)

An RCD is an electromechanical switching device whose contacts open when a residual current reaches a pre-determined level.

Its principle of operation is based on alternating residual magnetism, caused by an earth fault, being detected by a search coil which in turn operates a trip coil. When an equal current passes through line and neutral coils (i.e. everything is healthy) the alternating magnetism in the iron core, produced by these coils, will cancel out, leaving no residual magnetism.

When an earth fault occurs, the currents in the line and neutral coils are unequal because some current flows to earth, and these out-of-balance currents cause the alternating residual magnetic field required to operate the tripping mechanism.

The diagrams illustrate single and three phase RCDs.

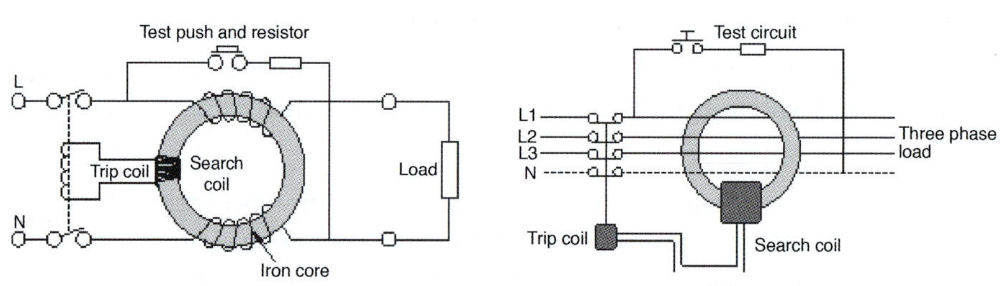

Fig. 19a Single phase RCD

Fig. 19b Three phase RCD

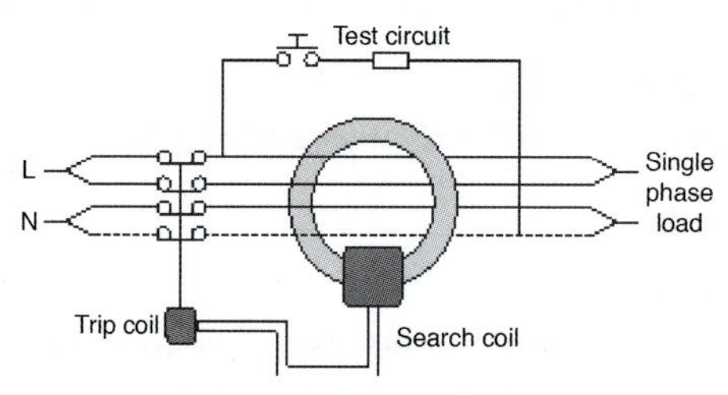

Fig. 19c Three phase RCD converted to single phase

RCD types

There are three categories of RCD:

1. Type AC......RCDs which can detect full wave a.c. residual currents only.
2. Type A........ RCDs which can detect full wave a.c. and pulsating d.c. residual currents.
   (Pulsating d.c. fault currents may be produced by loads that have rectifiers, thyristors, etc.)

3. Type B........RCDs that are able to detect full wave a.c., pulsating d.c. and pure d.c. residual currents.
   (Type Bs are not normally used in domestic installations)

Requirements for RCD protection

30mA

- All socket outlets rated at not more than 20 A and for unsupervised general use
- Mobile *equipment* rated at not more than 32 A for use outdoors
- All low voltage circuits in a bath/shower rooms
- Preferred for all circuits in a TT system
- All cables installed less than 50 mm from the surface of a wall or partition (even in the so-called safe zones) if the installation is un supervised, and also at any depth if the construction of the wall or partition includes metallic parts
- In zones 0, 1 and 2 of swimming pool locations
- All circuits in a location containing saunas, etc.
- Socket outlet final circuits not exceeding 32A in agricultural locations
- Circuits supplying Class II equipment in restrictive conductive locations
- Each socket outlet in caravan parks and marinas and final circuit for houseboats
- All socket outlet circuits rated not more than 32 A for show stands, etc.
- All socket outlet circuits rated not more than 32 A for construction sites (where reduced low voltage, etc is not used)
- All socket outlets supplying equipment outside mobile or transportable units
- All circuits in caravans
- All circuits in circuses, etc.

100mA

- Socket outlets of rating exceeding 32 A in agricultural locations

300mA

- At the origin of a temporary supply to exhibitions, shows, circuses, etc.
- Where there is a risk of fire due to storage of combustible materials
- All circuits (except socket outlets) in agricultural locations

500mA

- Any circuit supplying one or more socket outlets of rating exceeding 32 A on a construction site.

There are often circumstances occurring that cause unwanted tripping of RCDs; this topic is discussed under *Nuisance tripping*.

Residual current monitors (RCMs)

These devices are similar to insulation resistance monitors (IMDs), but in this case they continuously monitor the leakage currents in earthing systems and provide alarms when pre-determined levels of leakage are detected. Like IMDs they are used where disconnection of the supply is undesirable or not permitted.

Resistance R ohms

This is an opposition to the flow of current in a circuit.

Resistance *decreases* as the cross-sectional area of a conductor *increases*, and *increases* as the length *increases*. Hence, a 1.0 mm^2 conductor has 10 times more resistance than a 10.0 mm^2, and a 10 m length of conductor has 10 times less than 100 m.

Resistance also increases with a rise in temperature *(see also Temperature coefficient and Earth fault loop impedance test).* However, for materials such as silicon, germanium, carbon and battery electrolyte, the resistance *increases* with a *decrease* in temperature.

That is why it is often more difficult to start your car in very cold weather – the internal resistance of the battery increases at lower temperatures, which restricts the starting current.

Resistances may be connected in 'series', 'parallel', a combination of both or in more complex circuits, such as 'star, or 'delta' configurations.

In general, electrical installations, circuits and loads are connected in parallel, and hence the more circuits there are, the more cable there is connected in parallel. This allows more paths for current to flow between conductors through the insulation. This in turn results in an overall lowering of total insulation resistance *(see also Testing/insulation resistance).*

The topic of resistance in parallel always seems to be a cause of confusion. The total resistance of a number of resistances in parallel is found from:

$$\frac{1}{R_T} = \frac{1}{R_1} + \frac{1}{R_2} + \frac{1}{R_3} + \frac{1}{R_4} \text{......etc}$$

Example: The insulation resistances of four circuits are 20 M Ω, 30 M Ω, 100 M Ω and 50 M Ω.

R

The total resistance would be found from:

$$\frac{1}{R_T} = \frac{1}{20} + \frac{1}{30} + \frac{1}{100} + \frac{1}{50}$$

$$\text{So } \frac{1}{R_T} = 0.05 + 0.0333 + 0.01 + 0.02$$

$$\text{So } \frac{1}{R_T} = 0.1133$$

$$\text{So } R_T = \frac{1}{0.1133} = 8.83\text{M}\,\Omega$$

Note: For resistances in parallel the total resistance is always lower than the lowest value.

Resistivity ρ **($\mu\Omega$ mm)**

This is the resistance measured across the opposite faces of a unit cube of a material. Such a measurement is very low, in the order of millionths of an ohm, and the cube has 1 mm sides. Hence the units of measurement are $\mu\Omega$ mm.

Sometimes known as specific resistance, it indicates a standard resistance for materials, such that, given its resistivity, the resistance of any length or size of a material may be determined.

So the resistance **R** of a sample of material may be determined from its 'resistivity **ρ**' 'length **l**' and 'cross sectional area **a**'.

$$R = \frac{\rho \cdot L}{a} \text{ ohms}$$

Internationally, resistivity is measured and quoted at 20 °C.

The resistivity of pure copper is 17.2 $\mu\Omega$ mm. The copper used in electrical cables undergoes various changes during the manufacturing process and hence its resistivity is likely to be higher, in the region of 18 $\mu\Omega$ mm at 20 °C.

RIDDOR

The Reporting of Injuries, Diseases and Dangerous Occurrences Regulations 1995.

This requires employers to report injuries, diseases, etc.

Ring final circuits (see also Circuits)

Nearly all of us are familiar with ring circuits, and it is interesting to note that such systems are not peculiar to the UK; they are also used in the Republic of Ireland, Botswana, Ghana, Cyprus, Malta, Hong Kong, Macao, Oman, Kenya, Iraq and Qatar.

Appendix 15 of BS 7671:2008 gives pictorial details.

Risk assessment

This should be carried out before the commencement of an installation to ensure safety, not only during the installation process but also during the life of the system. The following list, which is not exhaustive, indicates some areas to be considered and could include, for example:

- The need for appropriate access equipment, ladders, scaffolding, etc.
- Provision of suitable Personal Protective Equipment (PPE), hard hats, safety shoes, etc.
- Compatibility of equipment
- The capability of the installation to be inspected and tested safely
- The choice of wiring systems to be appropriate for the environmental conditions.

r.m.s. values (see a.c.)

S

Safe isolation of supplies

The Electricity at Work Regulations 1989 prohibit, except under exceptional circumstances, live working. Hence, circuits that are to be worked on must be safely isolated.

The procedure is as follows:

1. Identify the circuit to be worked on.
2. Isolate the circuit and lock in the off position (in the case of fuses, remove them and keep on one's person or locked away).
3. Select an approved voltage indicator/test lamp conforming to GS38.
4. Prove the instrument on a known supply (same voltage as isolated circuit).
5. Test that the circuit is dead.
6. Re-prove the instrument.

Proving the instrument may be achieved using a 'proving unit' which delivers 230 V d.c. electronically.

(see also Test lamp)

Saunas

(BS 7671:2008 Section 703) The Regulations in this section are applicable to sauna cabins in rooms or sauna heaters, etc. in rooms, in which case the whole room is taken to be the sauna.

The diagram below indicates how the sauna is divided into zones.

Main points:

- All circuits of the location shall be protected by RCDs of maximum rating 30 mA. However the sauna heater need not be RCD protected unless recommended by the manufacturer.

The Dictionary of Electrical Installation Work. DOI: 10.1016/B978-0-08-096937-4.00018-0

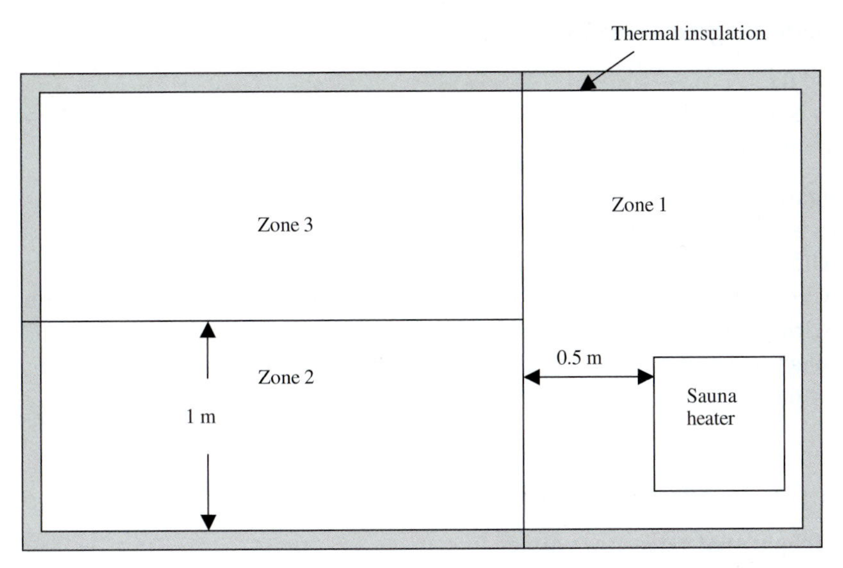

Fig. 20

- Equipment shall be to a minimum of IPX4 or IPX5 where water jets are used for cleaning
- In zone 1, only the sauna heater and associated equipment may be installed
- In zone 2 there are no restrictions regarding heat resistance of equipment as, of course, this is the cool area
- In zone 3 the temperature withstand for equipment is a minimum of 125 °C and that for cable insulation and sheaths is 170 °C
- It is preferred that wiring is installed outside of any of the zones. If it is installed in zones 1 or 3 it will need to be heat-resistant
- Sauna heater equipment switchgear may be installed in zone 2 in accordance with manufacturers' instructions. Other switches or controls, e.g. light switches, must be installed outside the sauna
- Socket outlets shall not be installed within the sauna heater location.

S

Schedule of inspections

This is a list of items checked during the inspection process. Appended against each item will be either a × (defect or omission) or a ✓ (satisfactory) or n/a (not applicable).

The × or the n/a may only be used on schedules associated with periodic inspections.

(see also Certification)

Schedule of test results

This schedule includes technical details of the supply, the method of fault protection and the instrument test results.

(see also Certification)

Schematic diagrams (see Diagrams)

Second fix

This is the installation of accessories and equipment after the first fix of the wiring system has been completed,

Segregation

This is the process whereby:

- Different currents and voltages in an enclosure are kept separate
- Circuits of different voltage bands are permitted to be installed in the same wiring system
- Electrical and non-electrical services do not have adverse effects on one another.

(see also Mutual detrimental influence)

Selectivity (see Discrimination)

SELV (separated extra-low voltage)

(BS 7671:2008 definition) 'An extra-low voltage system that is electrically separated from earth and from other systems in such a way that a single-fault cannot give rise to an electric shock.'

Although this system can be supplied from a motor-generator, a battery or certain electronic devices, it is more common to use a step-down isolating transformer. As the voltage is less than 50 V a.c. and there are no earths, this system provides both basic and fault Protection.

This system is generally found in locations where there is an increased risk of electric shock, such as bathrooms, swimming pools, etc.

(see also PELV)

Shock (see Electric shock)

Short-circuit current

(BS 7671:2008 definition) 'An overcurrent resulting from a fault of negligible impedance between live conductors having a difference in potential under normal operating conditions.'

In other words, a 'dead short' between brown, black or grey line conductors or between any of these and blue (neutral) *(see also Overcurrent).*

Simultaneously accessible parts

(BS 7671:2008 definition) 'Conductors or conductive parts which can be touched simultaneously by a person or, in locations specifically intended for them, by livestock.'

Such parts may be live parts or exposed and extraneous conductive parts

Skilled person

(BS 7671:2008 definition) 'A person with technical knowledge or experience to enable him/her to avoid the dangers that electricity may create.'

Clearly, electricians, electrical engineers, etc. fit the bill here.

Smoke detectors/alarm systems

There are two types: the 'photoelectric/optical' and the 'ionization'. The most common and inexpensive is the 'ionization' type, which contains a tiny (harmless) amount of radioactive material which emits a flow of electrons. When this flow is interrupted by smoke, an alarm is activated. (We don't need to go into nuclear physics here!.) This type is most suited to areas such as dining and living rooms, etc.

The 'photoelectric' variety relies on the interruption of an infra-red beam by smoke and is best installed in hallways and landings. These units may be 'stand-alone' or connected to a smoke alarm system.

Part 'B' of the Building Regulations details the requirements for new dwellings and extensions to existing ones. They require mains supplied systems **not** 'stand-alone' types.

Although it is highly recommended, there is no requirement to install smoke alarm systems where re-wiring has taken place.

Soil resistivity

This is the resistance per unit cube of soil and will vary depending on the soil type. The table below illustrates some typical average values in Ohms-m *(see also Resistivity)*

Soil type	Average resistivity Ωm
Marshy ground	2 – 2.7
Loam and clay	4 – 150
Sand	90 – 8000
Peat	200 upwards
Sandy gravel	300 – 500
Chalk	400 – 600
Rock	1000 upwards

The large range of values given for some of these soil types reflects the range in possible moisture content.

Solar gain

This is the increase in temperature of a space, material or structure resulting from solar radiation. The solar gain will increase as the strength of the sun increases.

Solar photovoltaic supply (see PV systems)

Space factor

This is a factor that is used to determine the number of cables which are permitted in trunking. The factor is 45%; in other words, cable should only occupy 45% of the space within the trunking.

It is generally accepted that the factor for conduit is 40%. However, it is not often that space factors would be used in practice, as there are sets of tables available giving cable sizes and associated trunking and conduit sizes.

Space heaters

As the name suggests these are appliances that heat a space! They may be thermal storage, radiant or convection types.

SPD (surge protection device) (see Surge protection)

Special locations

The BS 7671:2008 Part 7 lists 16 special locations:

- Bath and shower rooms
- Swimming pools
- Saunas
- Construction sites
- Agricultural locations
- Conducting locations with restricted movement
- Caravan/camping parks
- Marinas
- Medical locations
- Exhibitions, shows and stands
- Solar photovoltaic systems
- Mobile/transportable units
- Electrical installations in caravans
- Operating or maintenance gangways
- Fairgrounds, amusement parks and circuses
- Floor and ceiling heating systems

These are covered briefly in this book under their individual headings.

Split-load distribution board

These are distribution boards that are divided into sections, so that various parts of an installation may be controlled/protected individually. Typical of this arrangement is where only some circuits in an installation need to be RCD protected.

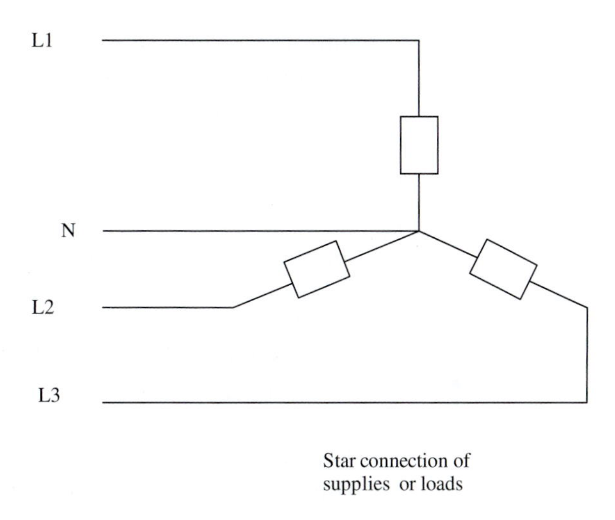

Fig. 21

Split load boards may be arranged in more than two sections, for example, where some circuits need additional protection by 30mA RCDs; others need 300mA protection against fire, and yet others need only overcurrent protection. In this case a three way split board would be used.

Spur

(BS 7671:2008 definition) 'A branch from a ring or radial circuit.'

Often misunderstood, a spur is the circuit, **not** the accessory.

Star connection

This is one way that three phase supplies or loads may be arranged. It is usual for the supply to an installation to be Star connected as there is a neutral (Fig. 21).

Three phase motors are seldom connected in Star except in the start-up mode for large motors with heavy starting currents *(see also Delta connections).*

Starter (fluorescent)

There are generally three methods of starting fluorescent lighting: the thermal starter, the quick or instant start and the glow starter.

The thermal starter comprises a pair of closed bi-metal contacts and a small heater in parallel with them. When the supply is switched on, the tube elements are energized via the contacts, and the heater also starts to warm up. This warming causes the contacts to open, thus open-circuiting the ballast in the luminaire, which in turn causes a large voltage to appear across the ends of the tube which then 'strikes' and illuminates.

The quick or instant starter is not really a starter as such; it is more of a method of getting the gas in the tube to ionize. This relies on the luminaire having an

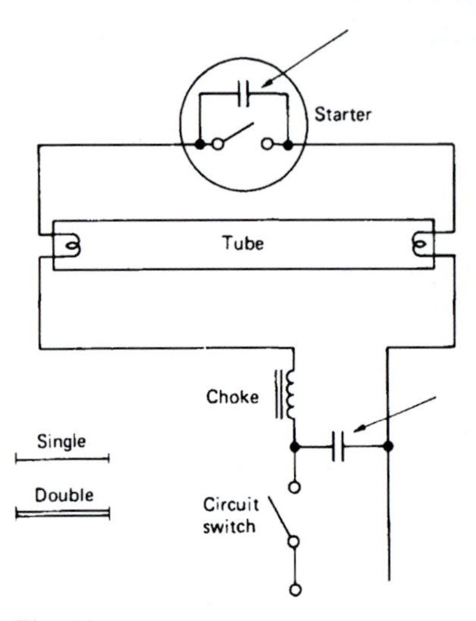

Fig. 22

auto-transformer and an earthed metal strip in close proximity to the tube. The starting process is achieved due to a difference in potential between the elements at the tube ends and the earthed strip, which causes ionization to spread along the tube.

The glow starter is the most common type. It works on the same principle as the thermal type. It comprises a pair of open bi-metal contacts enclosed in a sealed glass envelope filled with argon gas. When the luminaire is energized, a small current to the tube elements flows through the argon gas, warming it up – which cause the open contacts to close. This allows the full current to flow through the contacts and not the gas. The gas cools down and the contacts part, thus open-circuiting the ballast and causing a high voltage to appear across the tube ends, which ionizes/strikes the gas in the tube (Fig. 22).

Both the thermal and glow starters have an integral capacitor for radio interference suppression.

Starter (motor)

There are two general starter types: the Direct-on-Line (DOL) and the Star/Delta. Both rely on the 'hold-on' circuit principle *(see Hold-on circuit)*.

The DOL type are used for small motors, single or three phase, with light starting currents, whereas large three phase motors with heavy starting currents require the Star/Delta type.

The Star/Delta variety require a motor to have six terminals (the ends of the three windings). The windings are connected, via the starter, in Star which restricts the current, at start up, to a low value.

When the motor has reached a suitable speed the windings are automatically changed to the Delta configuration.

Starting current

This is the current drawn from the supply when items of equipment are switched on. It is associated with items such as motors, transformers, inductors (ballasts), etc.

This current, which does not last, is not usually taken into account when determining cable sizes.

It should, however, be considered where motors are subject to frequent starting and stopping and/or intermittent use, as temperature rise in equipment and cables may, over time, result in damage.

Stationary equipment

(BS 7671:2008 definition) 'Electrical equipment which is either fixed, or equipment having a mass exceeding 18 kg and not provided with a carrying handle.'

Strappers

These are the line conductors wired between switches in two way, or two way and intermediate switching systems.

Where these conductors are those incorporated in multi-core cables and coloured grey and black, they must be identified brown.

Street furniture

This is electrical equipment such as highway signs, bollards, lamposts, etc. *Not tables and chairs outside bistros!*

Stroboscopic effects

Fluorescent lighting often causes a flickering which can result in adjacent rotating machinery appearing stationary. Clearly this is a dangerous situation. It can be resolved by using 'lead-lag' circuitry which is achieved by wiring a capacitor in series with each alternate luminaire. This has the effect of creating opposing power factors which, in turn, cancel out the resultant flicker.

Another method for three phase systems is to spread the luminaires across the phases.

Supplementary equipotential bonding

This is used for 'additional protection' in situations where there is an increased risk of electric shock in such areas as bathrooms *(if applicable, see Bathrooms)*, swimming pools, agricultural locations, etc.

Supplementary equipotential bonding conductors connect together:

1. Two or more exposed conductive parts, or
2. Exposed conductive parts to extraneous conductive parts, or
3. Two or more extraneous conductive parts.

For items 1 and 2, the bonding conductor, if sheathed or mechanically protected, should have a conductance no less than **half** that of the cpc connected to the exposed conductive part. If not mechanically protected it should be a minimum of 4.0mm^2.

For item 3, the minimum sizes are 2.5mm^2 if mechanically protected and 4.0 mm^2 if not.

Supplementary insulation

(BS 7671:2008 definition) 'Independent insulation applied to basic insulation for fault protection.'

This provides for one the means of providing double insulation *(see also Class II equipment and Reinforced insulation).*

Surge protection

Surges are over-voltages caused by atmospheric events (lightning) or switching within an installation. Both may cause significant damage to installations and equipment. Protection against such over-voltages is achieved by surge protection devices (SPDs) which limit or divert these unwanted voltages.

Swimming pools

(BS 7671:2008 Section 702) A special location which includes pools, fountains, basins, shutes, flumes, etc.

As with bathrooms and saunas, swimming pools are divided into three zones:
Zone 0... the interior of the pool, basin or fountain
Zone 1... 2 m from the edge of zone 0 and 2.5 m above floor level
Zone 2... 1.5 m from the boundary of zone 1 and 2.5 m above floor level (there is no zone 2 for fountains).

Main points:

- Protection, dependent on the zone, may be provided by either SELV, automatic disconnection of supply together with 30mA RCDs, or electrical separation
- External influences are:

Zone 0....IPX8
Zone 1....IPX4 or IPX5 where cleaning is by water jets
Zone 2....IPX2 for indoor pools, IPX4 for outdoor pools or IPX5 where cleaning is by water jets

- Supplementary equipotential bonding is required between **all** extraneous and exposed conductive parts located in any of the zones
- Socket outlets and switches are only permitted in Zone 1 if they have a non-conductive protective covering, are located beyond 1.25 m from the boundary of Zone 0 and are mounted 300 mm above floor level.

S

Switch fuse

This is a switch where the fuse/s are adjacent to, and not part of, the switch blades.

(see also Fuse-switch)

Switched-mode power supply (SMPS)

An SMPS is an electronic power supply used largely in computers and associated equipment and provides a regulated voltage. Such devices can cause harmonic distortion problems.

(see Harmonic currents)

SY cable

This is flexible cable used in instrumentation, control, data applications, etc. It is multi-core, ranging from 2 to 60 core, and it is easily recognized as the outer sheath is transparent and the galvanized metal screen is visible.

System

The Electricity at Work Regulations 1989 (EAWR) define electrical **systems** and equipment as 'everything that produces, transfers, stores or uses electrical energy and all cables, conductors and equipment etc, that is or can be connected to the source of energy, including portable equipment and measuring devices.'

In other words, anything electrical, from power stations to batteries.

(see also Earthing systems)

T

Temperature coefficient α (units: $\Omega/\Omega/°C$)

This indicates the change in resistance of a 1Ω resistor of a material for every 1 °C rise in temperature.

If we take a 1Ω sample of copper at 0 °C, and increase its temperature by 1 °C, its resistance will increase by 0.004Ω to 1.004Ω. The value $0.004\,\Omega/\Omega/°C$ is the temperature coefficient, α, of copper.

Every material has a different coefficient, but the most common conducting materials have values ranging between 0.0039 and $0.0045\,\Omega/\Omega/°C$.

To determine the value of resistance of a material after a rise in temperature from 0 °C:

$$R_f = R_0(1 + \alpha t)$$

Where R_f = final resistance

R_0 = resistance at 0 °C

α = temperature coefficient

t = rise in temperature

For changes in temperature between any two values:

$$R_f = \frac{R_L(1 + \alpha t_f)}{(1 + \alpha t_L)}$$

Where R_f = final resistance

R_L = lower resistance

α = temperature coefficient

The Dictionary of Electrical Installation Work. DOI: 10.1016/B978-0-08-096937-4.00019-2

t_f = final temperature

t_L = lower temperature

Example: The resistance of a 1.0mm², 70 °C thermoplastic copper conductor is 18.1mΩ/m at 20 °C, its resistance at 70 °C will be:

$$R_f = \frac{18.1(1 + 0.004 \times 70)}{(1 + 0.004 \times 20)} = 21.45\text{m}\,\Omega$$

Which is approximately **1.2** times greater than the original 18.1mΩ.

This is the magical figure used to correct the 20 °C tabulated (R1 + R2) values to those for 70 °C thermoplastic copper cables.

(See also Resistance, R1 + R2, and Earth fault loop impedance)

Testing

This is carried out after the inspection of an installation and comprises, where relevant, the following:

1. Continuity of protective conductors including main and supplementary bonding conductors.
2. Continuity of ring final circuit conductors.
3. Insulation resistance.
4. Polarity.
5. Earth electrode resistance.
6. Earth fault loop impedance (external, Z_e and total, Z_s).
7. Additional protection by RCDs.
8. Prospective fault current.
9. Verification of phase sequence.
10. Functional testing.
11. Verification of voltage drop.

The BS 7671:2008 requires that tests 1 to 5 are carried out in that order before the installation is energized. However, it is important to conduct a loop impedance test **before** an RCD test as the RCD test places an earth fault on the system which will not be detected if there is no earth return path.

Continuity of protective conductors

Conducted with a ***Low Resistance Ohmmeter*** with a no-load voltage of between 4 V and 24 V d.c. or a.c. and a short circuit current of not less than 200mA.

Protective bonding conductors should be disconnected at one end, to avoid misleading results due to parallel resistance paths, and supplies should be isolated. Instrument test leads need to be zeroed.

Circuit protective conductors may be measured end-to-end but preferably linked to the line conductor at one end and measured between line and cpc at the other end, thus giving a value of (R1 + R2) for the circuit.

Continuity of ring final circuit conductors

Conducted with a ***Low Resistance Ohmmeter*** in order to identify any interconnections in the ring and to confirm the integrity of the cpc.

The individual L, N and cpc loops are measured.

Cross connect opposite Ls and Ns and measure between L and N at each socket. The readings should be substantially the same, and approximately ½ of the individual L and N values.

This indicates that there are no interconnections

Repeat this process but with opposite Ls and cpcs cross connected. For rings wired in flat twin with earth cable, the cpc is smaller than the line, and the readings at each socket will slightly increase and then decrease again as the test progresses round the ring. The highest reading will be approximately ¼ of the individual L and cpc values added together. This value is also the (R1 + R2) value for the ring.

This test method also automatically checks polarity at each socket.

Insulation resistance

Conducted using an ***Insulation Resistance Tester***, not a multi-meter, which is incapable of delivering the test voltages shown in the table below.

Test between live conductors, individually or collectively, to earth and between each live conductor.

The minimum values and the test voltages are:

Circuit voltage V	Test voltage d.c.	Minimum insulation resistance MΩ
SELV and PELV	250	0.5
LV up to 500 V and FELV	500	1.0
Over 500 V	1000	1.0

T

Where surge protective devices are present they should be disconnected. If this is not practicable, the test voltage may be reduced to 250 V, but the minimum insulation resistance remains at 1.0 MΩ.

Where the circuit contains electronic devices such as dimmer switches, the test should be conducted between the live conductors connected together and earth.

There are some very specialized locations in, for instance, hospitals and ammunition stores that are designated non-conducting locations. These require special insulation tests to be conducted between floors and walls, and need particular equipment for such tests. Needless to say such locations and the consequent test requirements are rare.

Polarity

Conducted using a ***Low Resistance Ohmmeter*** to ensure that:

- All single pole devices, switches, fuses, cbs, etc., are in the line conductor only
- Edison screw lampholders have the screwed part connected to the neutral. This does not apply to modern E14 and E27 lampholders, as these have insulating material for the thread
- All socket outlets and accessories are correctly connected.

Earth electrode resistance

This test is conducted using either an ***Earth Electrode Resistance Tester*** or an ***Earth Fault Loop Impedance Tester*** to determine the value of the resistance between the electrode and the earth.

The Earth Electrode Resistance Tester (battery operated) is either a three or four lead instrument, which is connected to the electrode under test and a current electrode, with a potential electrode placed midway between the two. A resistance measurement is taken, and then repeated twice more with the potential electrode placed at least 6 m either side. The average of the three readings is taken to be the resistance of the earth electrode.

(see also Earth electrode resistance area)

The ***Earth Fault Loop Impedance Tester*** is used when there is a supply protected by an RCD.

The standard earth fault loop impedance test for Z_e is carried out.

Earth fault loop impedance

This test is conducted using an ***Earth Fault Loop Impedance Tester*** to establish that an earth return path exists and is low enough to ensure operation of a protective device in the required time.

There are two tests, one for external loop impedance Z_e and one for total loop impedance Z_s. For the measurement of Z_e, the installation is isolated from the supply and the earthing conductor disconnected to avoid parallel paths. The instrument 'fly-leads' are connected to the incoming line neutral and earthing conductors.

When operated, the instrument connects a resistor (approx. 10 Ω) across L and E. This causes a current of around 23A to flow around the earth loop for a very brief time; typically less than 40 ms. The instrument electronics then carry out a calculation to give the Z_e value.

For Z_s, all earthing and bonding must be in place and the test carried out at the farthest point in each circuit. Clearly, placing a resistor between L and E creates an earth fault, if only for a very brief time, and hence RCDs protecting the circuit may operate before a reading can be taken. In these circumstances a value of Z_s should be determined by calculation from:

$$Z_s = Z_e + (R1 + R2)$$

T

On no account should the RCD be bridged out to enable the test to be conducted.

NOTE: Measured values of Z_s should *not* exceed 0.8 times the tabulated maximum values.

Additional protection by RCD

Conducted with an ***RCD Tester*** to establish the operating time of an RCD.

The instrument test leads L, N and E are connected to the appropriate terminals of the RCD or a circuit accessory preferably as close to the RCD as possible and the tester selected to the RCD rating (30mA or less). The tests and maximum operating times are:

½ $I_{\Delta n}$...........Should not trip
1 × $I_{\Delta n}$....... 300 ms
5 × $I_{\Delta n}$........40 ms

These tests should be conducted on both 0° and 180° settings and the longest operating time recorded.

Prospective fault current (PFC)

This test is conducted with a ***Prospective Fault Current Tester*** to ascertain the level fault current at all relevant points in an installation. The results of this test are used to verify that protective devices have the correct breaking capacity.

The test is carried out in a similar way as an earth fault loop impedance test. However two tests are required:

1. Prospective Earth Fault Current, PEFC, and
2. Prospective Short Circuit Current, PSCC.

The first is carried out between line and earth, the second, between line and neutral.

To determine the PSCC between lines of a three phase system, the single phase value is multiplied by 1.732, or 2 for an approximate value.

Check on phase sequence

This is conducted with either a rotating disc instrument or a lamp indicator. The purpose is to ensure that the phase sequence at the origin is maintained at distribution boards and motors throughout the installation.

Functional testing

In this case a check is made on equipment to ensure that it is properly mounted, installed and adjusted. In other words, does, for example, two way switching work properly, are heating controls correctly set, etc?

Where RCDs are used for fault and/or additional protection the test button should operated.

Verification of voltage drop

The BS 7671:2008 suggests that this verification is not normally needed for an initial verification, which implies that it is only required for an Electrical Installation Condition Report.

It further suggests that it be evaluated by:

1. Calculations using diagrams and charts which require the knowledge of cable lengths, load currents, etc. (quite complicated!), or
2. By measuring the circuit impedance (R1 + Rn) and multiplying by the design current I_b and a factor that adjusts for conductor operating temperature *(see Temperature coefficient)* usually 1.2 for 70 °C thermoplastic cable.

Whatever the method, for 400/230V supplies, the value obtained should not exceed the following:

Lighting 3%	230 V Single phase----6.9 V 400 V Three phase-----12 V
Other circuits 5%	230 V Single phase---- 11.5 V 400 V Three phase----20 V

Test lamps

These need to conform to the requirements of the HSE Guidance note GS38.

'Testascopes/neon screwdrivers', 'voltsticks' and the like are ***not*** approved voltage indicators and should ***never*** be used other than to give a vague indication of the presence or lack of voltage.

An approved test lamp should have the following attributes:

- Robust construction
- An indication of the level of voltage
- Double insulated leads
- Finger guards on the probes
- Fused probes
- Maximum of 2 mm exposed metal on probe tips.

(see also Safe isolation of supplies)

Thermal insulation

This comes in many forms, including, most commonly:

- Glass wool blankets
- Polystyrene
- Foam fill
- Reflective sheets
- Thermal blocks and plaster.

There are of course many more and, whilst they are so important for reducing energy loss, they do present a problem regarding the current-carrying capacity of cables. This problem is recognized and addressed in BS 7671:2008 where details of cable de-rating are given.

Installation reference method '**A**' is where non-sheathed cables in conduit or multi-core cables are installed where they are in contact with thermal insulation on one side only, for example, in a stud wall or above a ceiling where the cable/conduit is touching the inner wall/ceiling surface.

The tables in Appendix 4 of BS 7671:2008 give current-carrying capacities for cables installed to method '**A**' which are in the region of 25% to 30% less than those for cables that are 'open or clipped direct' (reference method '**C**').

Method 100
Touching insulation one side

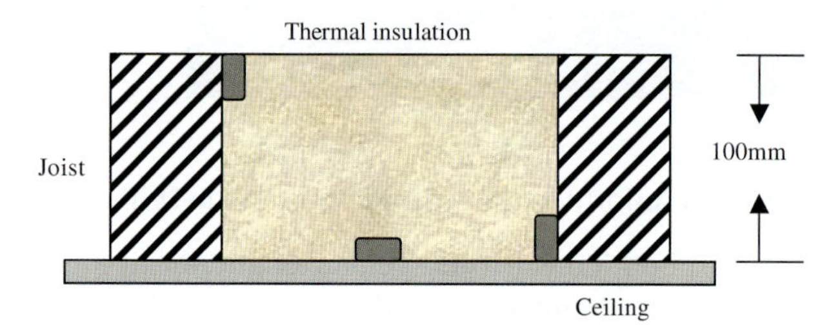

Fig. 23a

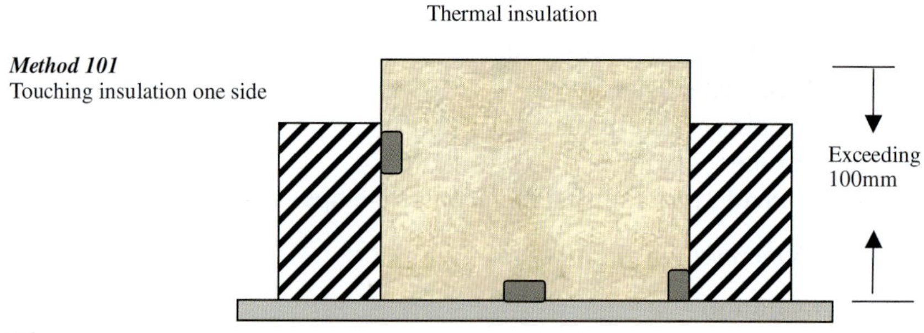

Method 101
Touching insulation one side

Fig. 23b

T

Method 102
Touching inner wall surface

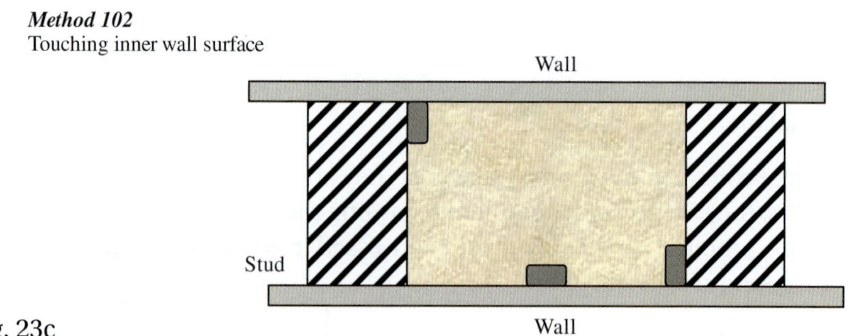

Fig. 23c

Method 103
Not touching inner wall surface

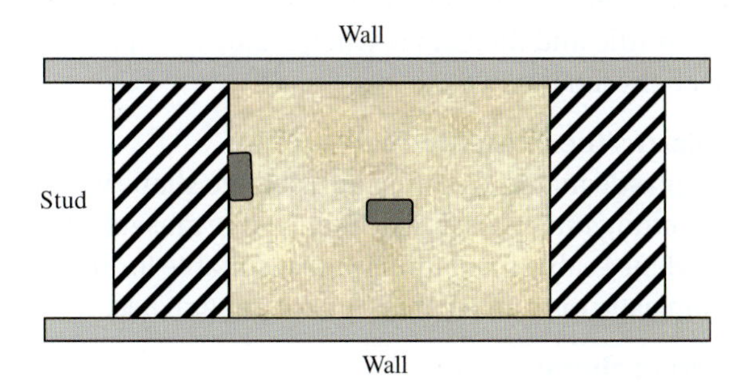

Fig. 23d

Method A
Single core cables in conduct or multi-core cables in an insulated wall

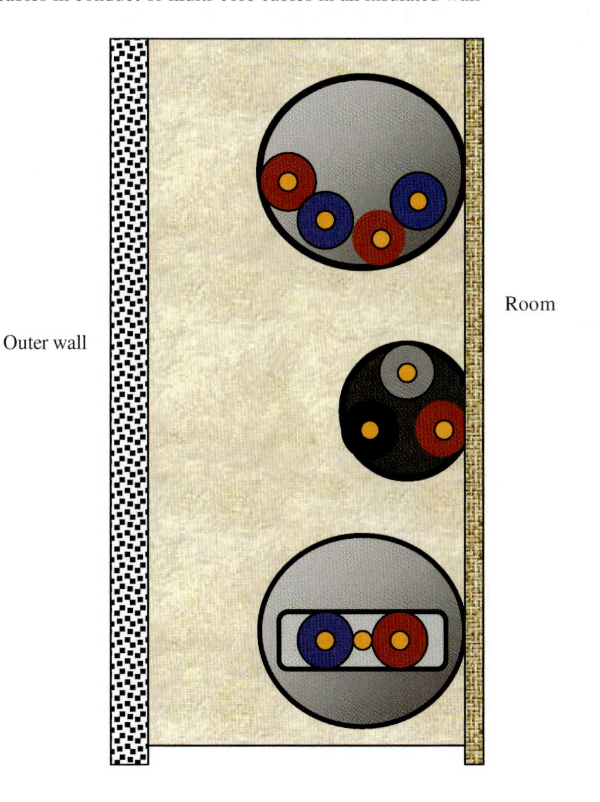

Fig. 24

For flat sheathed cables, Table 4D5 of the BS 7671:2008 gives values for five reference methods involving thermal insulation; 100, 101, 102 103 and '**A**.' These are shown in Figs 23a to 23d and 24.

It interesting to note that pvc in contact with polystyrene produces a chemical reaction that causes the plasticizer in the pvc to migrate into the polystyrene, and although the electrical qualities of the pvc are not diminished, it is likely to become brittle and then crack when disturbed. It is sensible, therefore, to avoid contact between the two.

Thermoplastic/thermosetting cable

Thermoplastic materials soften and change shape when heated whereas thermosetting materials do not. Hence thermosetting pvc (XLPE, cross-linked polyethylene) cable insulation is most suitable for use in high temperature areas. Thermoplastic insulated cables are for general use.

Time/current characteristics

These relate to fuses and circuit breakers and are sets of graphical curves produced by the protective device manufacturers and reproduced in BS 7671:2008.

They indicate the times 't' at which a particular device will operate when subjected to an overcurrent.

TN, TN-C-S and TN-S systems (see Earthing systems)

Transformer ratios

These are the ratios of current I, voltage V and number of turns N on the windings, between the primary and secondary sides of a transformer. These ratios can be shown as follows:

$$\frac{V_p}{V_s} = \frac{N_p}{N_s} = \frac{I_s}{I_p}$$

If we take the current and voltage ratios $V_p/V_s = I_s/I_p$ and cross multiply we get:

$$V_p \cdot I_p = V_s \cdot I_s$$

Which shows that, ignoring losses, the primary power is the same as the secondary power (remember power is voltage × amperes)

Transformer ratings are given in VA or kVA and, at a basic level, this means that a 20VA transformer will transfer 20VA of power from primary to secondary.

T

Hence, a 230V/12V, 250VA transformer that is used to supply 12 V extra-low voltage lamps would have a primary current Ip of 250/230 = 1.08A, and a secondary current Is of 250/12 = 20.8 A. So, the cable supplying the 12 V lamp will have to have a significantly larger cross-sectional area than that of the primary.

Transportable units (see Mobile and transportable units)

Trunking

Trunking is an enclosure or containment system which is used to minimize the risk of damage to cables. It may comprise a complete system or isolated lengths for cable drops or runs to accessories.

Trunking is available in a range of sizes, and is either metallic (galvanized or various paint finishes) or pvc. Compartmentalized varieties are available where segregation of circuits is a requirement.

Steel trunking may be used as a cpc although rarely in modern installations, where a separate cpc is provided. Steel trunking is an exposed conductive part, and hence needs to be continuous and connected to earth. This will be achieved if the system is installed correctly. Some installers go the extra mile and fit links between each join, but this is not really necessary. This is also the case with the 'fly-lead' from an accessory to a 'back-box' which is only required by BS7671:2008 where the trunking is used as the cpc.

The popular pvc min-trunking tends to be used as a cosmetic covering for flat sheathed cables rather than a containment system.

Trunking capacity

As an easily accessible containment system, it is often tempting to keep adding circuits in existing trunking. Great care should be taken if this is the intention, as trunking capacity is based on a 'space factor' of 45% and exceeding this may cause the existing cables to be under-rated *(see also Space factor)*.

It should be noted that Band I (ELV) and Band II (LV) circuits may need to be installed in trunking that provides segregation facilities.

(see also Segregation)

TT system (see Earthing systems)

U

U_0

This is the nominal voltage to earth; usually 230 V for most installations.

Underground cables

Cables installed underground should be buried at a depth to avoid mechanical damage. There are several Special Locations, identified in the BS 7671:2008, where specific depths are quoted, with or without additional mechanical protection. This additional protection can be provided by installing the cables in ducting, etc., or covering with cable tiles.

Undervoltage

This includes complete loss of supply, as well as the effects of overloads or disturbances on the DNOs distribution network, or starting currents etc. in an installation which may cause the voltage to drop.

This loss or drop may be of sufficient duration to de-energize motors. Under these circumstances it is important that restoration of the supply to its normal value should not be automatic. One does not want to be clutching a band-saw blade while investigating why it is not moving, when the supply automatically resumes. Ouch!

Fortunately most items of motor control equipment have a 'no-volt release' *(see also Hold-on circuits)* mechanism which means the equipment will need to be manually re-started.

Clearly there are circumstances where a loss or drop in voltage causing equipment malfunction is undesirable, e.g. life support systems, emergency lighting, fire rescue lifts, fire alarm systems, etc. In these cases alternative supplies, 'safety services', are needed. The sources for such services include, for example, batteries or generators.

The Dictionary of Electrical Installation Work. DOI: 10.1016/B978-0-08-096937-4.00019-2

UPS (uninterruptible power supply)

This is a battery back-up system where the d.c. output from the batteries is converted, electronically, to a.c. UPSs are used to provide an emergency source for safety services and are also used extensively where computers and data equipment rely on an uninterrupted supply.

V

Volt symbol V

This is the unit of electrical pressure and named after the Italian scientist Alessandro Volta (1745–1827).

Voltage bands

There are two bands or ranges of voltage stated in the BS 7671:2008 – Band I and Band II.

- **Band I (ELV)**...not exceeding 50V a.c. or 120 V ripple-free d.c. between conductors and between conductors and earth
- **Band II (LV)**...exceeding extra-low but not exceeding 1000V a.c. or 1500 V d.c. between conductors, or 600V a.c. or 900 V d.c. between conductors and earth.

There is no band for high voltage; BS 7671:2008: just states that HV normally exceeds LV.

(see also Extra-low voltage and Low voltage)

Voltage drop

When a current flows, a conductor's resistance causes a drop in voltage along its length. The magnitude of this 'volt-drop' will depend on the conductor material, its length, cross-sectional area and the value of the current flowing.

The BS 7671:2008 gives tabulated values of 'volt-drop' in milli-volts (mV), for every ampere of current (A), for every metre of length (m), for each size and type of cable. Hence:

$$\text{Volt-drop} = \frac{\text{mV} \times \text{A} \times \text{m}}{1000}$$

(the 1000 converts the mV to whole volts)

The Dictionary of Electrical Installation Work. DOI: 10.1016/B978-0-08-096937-4.00020-9

In some of the BS 7671:2008 tables the volt-drop is divided into three columns headed **r**, **x** and **z.** This occurs with conductors of 16.0 mm^2 and above.

At these sizes there is likely to be a resistive, **r**, and a reactive, **x**, volt-drop. The overall drop, **z**, is a combination of these two. It is generally only in more complex designs that there is a requirement to know the details of volt-drop due to **r** or **x**. Hence the **z** value is usually used *(see also Impedance)*.

The BS 7671:2008 gives maximum values of voltage drop in an installation, from the origin, of 3% for lighting and 5% for other circuits *(see also verification of voltage drop under **Testing**)*.

Voltage optimization

This is also known as 'voltage control' or 'voltage stabilization'.

Invariably, the voltage supplied by the DNOs is higher than that required by the equipment. Typically, modern electrical items are rated at 220V, whilst the supply voltage is permitted to vary between 216.2V and 253V. The majority of supplies, when measured, will be in the region of 240V.

Supplying 220V equipment at 240V causes inefficiency and losses and also shortens equipment life. By reducing the supply voltage to the optimum required for safe and efficient functioning, there will be a decrease in energy, thus reducing electricity bills and carbon emissions.

At a basic level, voltage optimization is achieved by use of a transformer unit with various voltage tappings which are changed to suit the loads. These units are usually controlled automatically as loads and supply voltages change and although, in the past, they have been restricted to large commercial and industrial premises, units are now being produced for domestic applications.

Warning notice

The BS 7671:2008 requires warning notices for the following:

- Voltages exceeding 230 V where such voltages are not normally expected
- Where live parts cannot be isolated by a single device
- At earthing connections at earth electrodes, extraneous conductive parts and the main earthing terminal when separate from main switchgear
- In earth-free equipotential zones where connections to earth are not permitted
- Non-standard colours
- Dual supply.

Water pipes

Public water utility pipes are **not** permitted to be used as earth electrodes. However, other water pipes are permitted to be used, provided precautions are taken against their removal and they have been considered for such use.

Watt symbol W

This is the unit of electrical power named after James Watt, a British engineer (1736–1819) who invented the improved steam engine and introduced 'horse-power', (h.p.), as a means of measuring power. There are 746 Watts in 1 hp.

(see also Horsepower)

Wiring diagrams (see Diagrams)

Work symbol W

When an object is moved from one place to another 'work' is done. The amount of work done is dependent on the force or weight (F) and the distance (l) travelled, hence:

The Dictionary of Electrical Installation Work. DOI: 10.1016/B978-0-08-096937-4.00021-0

$W = F \times l$ joules

Where W = work in joules (J)

F = load force in Newtons (N)

l = distance in metres (m)

Double insulation (see Class II equipment)
Duct
Duty-holder

Earth
Earth electrode
Earth electrode resistance
Earth electrode resistance area
Earth fault current
Earth fault loop impedance Zs (see also Earthing systems and Testing)
Earth leakage (see Protective conductor current)
Earthing
Earthing conductor
Earthing systems
ECA
ECS card (see Construction Skills Certification Scheme)
Eddy currents
ELECSA
Electric shock
Electrical installation certificate EIC (see Certification)
Electrical installation condition report EICR (see Certification)
Electrical separation
Electricity at Work Regulations (EAWR) 1989
Electrical, Safety, Quality and Continuity Regulations (ESQCR) 2002
Electromagnetic compatibility (EMC)
Electromotive force (e.m.f.)
Electron
Enclosure (see also Barriers)
Equipotential bonding
Exhibitions, shows and stands
Exposed conductive part
External influence (see also IP and IK codes, Barriers and Enclosures)
Extra low voltage, ELV (see also SELV, PELV, FELV and Band I)
Extraneous conductive part

Fairgrounds, amusement parks and circuses
Farad (symbol F)
Faraday cage (see Equipotential bonding)
Fault current I_f
Fault protection
FELV (functional extra-low voltage)
Ferromagnetic material
First fix
Flexible cables
Floor and ceiling heating systems

Appendix

GENERAL FORMULAE AND INFORMATION

Multiples of units

Prefix	Value	Symbol
tera	10^{12}	T
giga	10^{9}	G
mega	10^{6}	M
kilo	10^{3}	k
centi	10^{-2}	c
milli	10^{-3}	m
micro	10^{-6}	μ
nano	10^{-9}	n
pico	10^{-12}	P

Typical examples

1 million ohms = 1 megohm = 1MΩ
1 thousand watts = 1 kilowatt = 1kW
1 thousandth of an ampere = 1 milliampere = 1mA
1 millionth of a volt = 1 microvolt = 1μV
1 million millionth of a farad = 1 picofarad = 1pF

Note

For example: 10^{-6} may be shown as $\frac{1}{10^{6}}$ and $\frac{1}{10^{-6}}$ may be shown as 10^{6}

So $x \times 10^{-6}$ may be shown as $\frac{x}{10^{6}}$ and vice versa and

$\frac{x}{10^{-6}}$ may be shown as $x \times 10^{-6}$ and vice versa

Systeme International (SI units)

The base units are as follows.

Length	metre	m
Mass	kilogramme	kg
Time	second	s
Electric current	ampere	A
Luminous intensity	candela	cd
Thermodynamic temperature	kelvin	K
Amount of substance	mole	mol

Length

The units in general use and their conversions are as follows.
Millimetre (mm);
centimetre (cm);
metre (m);
kilometre (km).

To obtain	multiply	by
mm	cm M Km	10^1 10^3 10^6
Cm	mm m km	10^{-1} 10^2 10^5
M	mm cm km	10^{-3} 10^{-2} 10^3
Km	mm cm m	10^{-6} 10^{-5} 10^{-3}

Area

square millimetre (mm^2);
square centimetre (cm^2);
square metre (m^2);
square kilometre (km^2)
also, $1km^2 = 100$ hectares (ha)

To obtain	multiply	by
mm^2	cm^2 m^2 km^2	10^2 10^6 10^{12}
cm^2	mm^2 m^2 km^2	10^{-2} 10^4 10^{10}
m^2	mm^2 cm^2 km^2	10^{-6} 10^{-4} 10^6
km^2	mm^2 cm^2 m^2	10^{-12} 10^{-10} 10^{-6}

Volume

cubic millimetre (mm^3);
cubic centimetre (cm^3);
cubic metre (m^3)

To obtain	multiply	by
mm^3	cm^3 m^3	10^3 10^9
cm^3	mm^3 m^3	10^{-3} 10^6
m^3	mm^3 cm^3	10^{-9} 10^{-6}

Capacity

millilitres (ml);
centilitre (cl); litre (l);
also, 1 litre of water has a mass of 1 kg.

To obtain	multiply	by
ml	cl l	10^1 10^3
cl	ml l	10^{-1} 10^2
l	ml cl	10^{-3} 10^{-2}

Mass

milligramme (mg);
gramme (g);
kilogramme (kg); tonne (t).

To obtain	multiply	by
mg	g kg t	10^3 10^6 10^9
g	mg kg t	10^{-3} 10^3 10^6
kg	mg g t	10^{-6} 10^{-3} 10^3
t	mg g kg	10^{-9} 10^{-6} 10^{-3}

Temperature

Kelvin (K) = °C + 273.15
Celsius (°C) = K − 273.15

Celsius (°C) = $\frac{5}{9}$(°F − 32)

Fahrenheit (°F) = $\left(\frac{9 \times °C}{5}\right) + 32$

Boiling point of water at sea level = 100°C or 212°F
Freezing point of water at sea level = 0°C or 32°F
Normal body temperature = 98.4°F or 36.8°C

Areas of figures and solids

Square = $a \times a = a^2$ Rectangle = $h \times l$
Parallelogram = $h \times l$ Cube = $6a^2$ Cuboid = $2hb + 2hl + 2bl$ or $2(hb + hl + bl)$

Circle = πr^2 or $\frac{\pi d^2}{4}$ ($\pi = 3.1416$) Ellipse = πxy Sphere = $4\pi r^2$ Perimeter = $\pi(x + y)$

Circumference = $2\pi r$ or πd

Cone:
open ended = πrl
solid = $\pi rl + \pi r^2$
or $\pi r(l + r)$

Cylinder:
hollow = $2\pi rh$
one ended = $2\pi rh + \pi r^2$
or $\pi r(2h + r)$
solid = $2\pi rh + 2\pi r^2$
or $2\pi r(h + r)$

Volumes of solids

Cube = $a \times a \times a = a^3$ Cuboid = $h \times b \times l$

Sphere = $\frac{4}{3}\pi r^3$ Cone = $\frac{1}{3}\pi r^2 h$ Cylinder = $\pi r^2 h$

TRIGONOME T RICAL RATIOS

For a right-angled triangle:

$\sin\theta = \frac{B}{A}$, $\cos\theta = \frac{C}{A}$

$\tan\theta = \frac{B}{C}$

$\frac{\sin\theta}{\cos\theta} = \tan\theta$

$\sin^{-1}$ means 'The angle whose sine is ....'
So, $\sin^{-1} 0.5$ means, the angle whose sine is 0.5; also $\cos^{-1} 0.5$ means, the angle whose cosine is 0.5; and $\tan^{-1} 0.5$ means, the angle whose tangent is 0.5.

The theorem of Pythagoras

For a right-angled triangle:
'The square on the hypotenuse (A) is equal to the sum of the squares on the other two sides.'
Hence:
$A^2 = B^2 + C^2$
$\therefore A = \sqrt{(B^2 + C^2)}$
or $A = (B^2 + C^2)^{1/2}$ $\quad X^{1/2}$ means $\sqrt{x}$

Basic mechanical formulae

Force (F) in newtons	=	mass (m) in kg	× acceleration (a) in m/s^2
Work donein joules	=	force (F) in newtons	× distancem metres
Load force (F) in newtons	=	mass (m) in kg	× acceleration due to gravity (g) where $g = 9.81\ m/s^2$

Simple leavers

At balance $F \times l = E \times L$.

BASIC ELECTRICAL FORMULAE

Ohm's law

$$R = \frac{V}{I} \text{ ohms}$$

Where:
I = current in amperes
V = voltage in volts
R = resistance in ohms

Resistivity

$$R = \frac{\rho \cdot l}{a}$$

where:
ρ = resistivity in ohm metres
l = length in metres
a = cross – sectional area in square metres

Temperature coefficient

For rise in temperature from 0°C to t°C

$R_f = R_o(1 + \alpha t)$ ohms

where:

R_f = final resistance
R_o = resistance at 0°C
α = temperature coefficient
t°C = change in temperature
For change from t_1 to t_2

$$\frac{R_1}{R_2} = \frac{(1 + \alpha t_1)}{(1 + \alpha t_2)}$$

Resistances in series

$R_{total} = R_1 + R_2 + R_3$ etc.

Resistances in parallel

$$\frac{1}{R_{total}} = \frac{1}{R_1} + \frac{1}{R_2} + \frac{1}{R_3} \text{ etc.}$$

Power in resistive circuits

$P = I.V$ watts

or

$$P = \frac{V^2}{R} \text{ watts}$$

Electrical heating

Heat energy = mass × change in temperature × specific heat capacity (joules)

$$\text{kWh output} = \frac{\text{Mass} \times \text{change in temperature} \times \text{SII}}{3600000}$$

Quantity of electricity and charge on a capacitor

$Q = I.t$ coulombs

where:

Q = charge
I = current
t = time in seconds

Charge is also
$Q = C.V$ coulombs

where:

C = capacitance in farads
V = voltage across plates

Electric force of field strength

$$E = \frac{V}{d} \text{ volts/m}$$

where:
V = voltage across plates
d = distance between plates in metres

Electric flux density

$$D = \frac{Q}{a} \text{ coulombs/m}^2$$

Where:
Q = charge
a = area of plates in square metres

Absolute permittivity

$$\varepsilon = \frac{D}{E} \text{ farads/metre}$$

where:
E = electric force
D = electric flux density

$$\text{or } \varepsilon = \varepsilon_o \varepsilon_r$$

where: ε_r = relative permittivity of dielectric
ε_o = permittivity of free space
$= 8.85 \times 10^{-12}$ F/m

also $$C = \frac{\varepsilon_o \varepsilon_r A(n-1)}{d} \text{ farads}$$

where:
C = capacitance of capacitor
n = number of parallel plates
d = distance between plates in metres
A = area of each plate in square metres

Energy stored in a capacitor

$$W = \frac{1}{2} \cdot C \cdot V^2 \text{ joules}$$

where:
C = capacitance in farads
V = voltage across plates

Capacitors in parallel

$C_{total} = C_1 + C_2 + C_3$ etc.

Charge on each capacitor:

$Q1 = C_1 . V : Q_2 = C_2 . V : Q_3 . V$: etc.

Capacitors in series

$$\frac{1}{C_{total}} = \frac{1}{C_1} + \frac{1}{C_2} + \frac{1}{C_3} \text{etc.}$$

Charge on each capacitor is the same as the total charge:
$Q_1 = C . V : Q = C_1 . V_1 : Q = C_2 . V_2 :$ etc.

Time constant

$\tau = C . R$

where:
C = capacitance in farads
R = resistance in ohms

Magneto motive force (m.m.f.)

$F = NI$ ampere turns

where:
F = m.m.f
N = no. of turns
I = current in amperes

Magnetizing force

$H = \frac{NI}{l}$ ampere turns/metre

*w*here:
l = length of magnetic circuit

Absolute permeability

$$\mu = \frac{B}{H}$$

where: B = flux density in teslas
H = magnetizing force in ampere – turn/metre

or $\mu = \mu_0 . \mu_r$

where: μ_0 = permeability of free space = $4\pi \times 10^{-7}$H/m
μ_r = relative permeability of magnetic material

Magnetic flux

$$\Phi = \frac{F}{S}$$

where:
Φ = flux in webers
F = m.m.f.
S = reluctance

also $S = \dfrac{l}{\mu A}$

for reluctance in series:
$S = S_1 + S_2 + S_3$

Magnetic flux density

$B = \dfrac{\Phi}{A}$ teslas

where:
Φ = flux in webers
A = area at right angles to field

Force on a conductor

$F = B\,.\,l\,.\,I$ newtons
where:
B = flux density
l = length of conductor in metres
I = current

E.m.f. induced by conductor

$E = B\,.\,l\,.\,v$ volts
where:
B = flux density
l = length of conductor in metres
v = velocity in metres/second

For a conductor moving at an angle, then:

$E = B\,.\,l\,.\,v\,.\,\sin\theta$

where:
θ = angle in degrees

Induced e.m.f. due to flux change

$$E = \frac{(\Phi_2 - \Phi_1)}{t} . N \text{ volts}$$

where:
Φ = flux in webers
N = no. of turns
t = time in seconds

Self inductance

$$E = \frac{-L(I_2 - I_1)}{t} \text{ volts}$$

Note: The minus sign denotes the e.m.f. is a back e.m.f.

also $L = \dfrac{N \cdot \Phi}{I}$ henrys

where:
E = induced e.m.f.
L = self inductance in henrys
I = current
t = time in seconds

Mutual inductance

$$E = \frac{-M(I_2 - I_1)}{t} \text{ volts}$$

$$M = \frac{(\Phi_2 - \Phi_1)}{(I_2 - I_1)} \cdot N \text{ henrys}$$

Energy stored in a magnetic field

$$W = \frac{1}{2} \cdot L \cdot I^2 \text{ joules}$$

Time constant

$$\tau = \frac{L}{R}$$

ELECTRICAL MACHINES

D.c. motors

$$E = \frac{2 \cdot p \cdot \Phi \cdot n \cdot z}{c}$$

where:
E = back e.m.f
p = no. of pairs of poles
Φ = flux per pole
n = speed in rev/s
z = no. of armature conductors
$c = 2_p$ for lap wound and 2 for wave wound
also $E = V - I_a \cdot R_a$

where:
V = supply voltage
I_a = armature *current*
R_a = armature resistance

Torque equation

$$\frac{T_1}{T_2} = \frac{I_{a1} \cdot \Phi_1}{I_{a2} \cdot \Phi_2}$$

D.c. generators

$$E = \frac{2 \cdot p \cdot \Phi \cdot n \cdot z}{c}$$

and $E = V + I_a R_a$

$$E_1 = \frac{n_1 \cdot \Phi_1}{n_2 \cdot \Phi_2}$$

$E \propto n \,.\, \Phi$

where:
E = generated e.m.f.

Mechanical output

$P = 2 \cdot \pi \cdot n \cdot T$ watts

where:
n = speed in rev/s
T = torque in newton metres

Induction motors

$F = n \cdot p$

where:

f = frequency in hertz
n = speed in rev/s
p = no. of pairs of poles

Slip

Percentage slip $(s) = \frac{n_s - n_\tau}{n_s} \times 100$

Per unit slip $(s) = \frac{n_s - n_\tau}{s}$

where:
n_s = synchronous speed
n_τ = rotor speed

Slip speed $(s) = n_s\, n_\tau$
Slip frequency $f_s = sf$

where: f = supply frequency
s = slip

A.C. CIRCUITS

R.m.s. value = 0.707 × maximum value
Average value = 0.636 × maximum value

Inductive reactance

$X_1 = 2 \cdot \pi \cdot f \cdot L$ ohms
$V = I \,.\, X_L$ volts

where:
f = frequency in hertz
L = inductance in henrys

Inductance (L) and resistance (R) in series

$V_{X_L} = I \cdot X_L$ ohms
$V_R = I \cdot R$
$V = V_{X_L} + V_R$ by phasors only

or $V^2 = V_{X_L}^2 + V_R^2$ if components are assumed pure
Total opposition = impedance = Z ohms

$Z = \sqrt{R^2 + X_L^2}$ ohms
or $Z = \frac{V}{I}$

L and R in parallel

$V = I_2 \,.\, X_L$
or $V = I_1 \,.\, R$

$Z = \frac{V}{I}$

$I = I_1 + I_2$ by phasors or active and reactive components only
or $I^2 = I_1^2 + I_2^2$ if components are assumed pure

Power factor

Power factor (PF) = $\cos\theta$

or $= \frac{R}{Z}$

Capacitive reactance

$X_C = \frac{1}{2 \cdot \pi \cdot f \cdot C}$ ohms

where:
f = frequency in hertz
C = capacitance in farads
$V = I \times X_C$

Capacitance and resistance in series

$V_{X_C} = I \cdot X_C$

$V_R = I \cdot R$

$V = VX + VR$ by phasors only

or $V^2 = V_R^2 + V_{XC}^2$ if components are assumed pure

Impedance

$Z = \sqrt{R^2 + X_C^2}$ ohms

or $Z = \frac{V}{I}$

Capacitance and resistance in parallel

$V = I_2 \cdot X_c$
or $V = I_1 \cdot R$
or $V = I \cdot R$
$I = I_1 + I_2$ by phasors, or active and reactive components only
or $I^2 = I_1^2 + I_2^2$ if components are assumed pure

Inductance, capacitance and resistance in series

$V_X = I \cdot X_L$
$V_X = I \cdot X_C$
$V_R = I . R$
$V = I . Z$
$V = V_{X_L} + V_{X_c} + V_R$ by phasors or active and reactive components only

or $V^2 = V_R^2 + \left(V_{X_L} - V_{X_c}\right)^2$ if components are assumed pure

Impedance

$Z = \sqrt{\left[R^2 + X_L - X_C^{\ 2}\right]}$ ohms

or $Z = \dfrac{V}{I}$

Resonance

$f_o = \dfrac{1}{2\pi\sqrt{(L.C)}}$ hertz

Also at resonance

$X_L = X_C$

so $Z = \sqrt{R^2} = R$ ohms

Voltage magnification, dynamic impedance or Q factor

$Q = \dfrac{1}{R}\sqrt{\left[\dfrac{L}{C}\right]}$

POWER

Single phase

Power triangle:
Power factor = $\cos\theta = \dfrac{W}{Va}$
hense $W = Va \cos\theta$
and $Va = Va \cdot \sin\theta$

Three phase

Star connected:

$VL = \sqrt{(3)} \cdot V_p$

$I_L = I_P$

where:
V_L = line voltage
V_p = phase voltage
I_L = line current
I_p = *phase current*
Delta connection:
$V_L = V_P$
$IL = \sqrt{(3)} \cdot IP$
For star or delta connection:
Total $Va = \sqrt{(3)} \cdot V_L \cdot I_L$
and, total watts $= \sqrt{(3)} \cdot V_L \cdot I_L \cdot \cos\theta$

Delta ← → star conversions

To convert (1) to (2)

$$R_1 = \frac{R_{AB} \times R_{CA}}{R_{AB} + R_{BC} + R_{CA}}$$

$$R_2 = \frac{R_{AB} \times R_{BC}}{R_{AB} + R_{BC} + R_{CA}}$$

$$R_3 = \frac{R_{BC} \times R_{CA}}{R_{AB}+R_{BC}+R_{CA}}$$

To convert (2) to (1)

$$R_{AB} = R_1 + R_2 + \frac{(R_1 \times R_2)}{R_3}$$

$$R_{BC} = R_2 + R_3 + \frac{(R_2 \times R_2)}{R_1}$$

$$R_{CA} = R_3 + R_1 \frac{(R_3 \times R_1)}{R_2}$$

TRANSFORMERS

$$\frac{V_p}{V_s} = \frac{N_p}{N_s} = \frac{I_s}{I_p}$$

Regulation

$$\text{Percentage} \times \text{regulation} = \frac{E_s - V_s}{E_s} \times 100$$

Where: E_s = no load voltage
V_s = on load voltage

Losses

Hysteresis loss α frequency
Eddy current $\propto \Phi^2{}_{max} \times f^2 \times$ (*lamin* ation thickness)^2

Efficiency

$$\text{Percentage efficiency} = \frac{\text{output}}{\text{input}} \times 100$$

RECTIFIERS

Half wave:
Average value $= \frac{0.636 \times \text{max.value}}{2}$

WHEATSTONE BRIDGE

$$\frac{P}{Q} = \frac{R}{X}$$

CELLS AND BATTERIES

Internal e.m.f.

$$E = V + Ir$$

where:
E = e.m.f. of cell
V = terminal p.d of cell
I = current dawn
r = internal resistance of cell

or $E = V + I(R + r)$
where: R = resistance of connected load

Ampere hour (Ah) efficiency

$$\text{Ah efficiency (\%)} = \frac{\text{Ah output}}{\text{Ah input}} \times 100$$

Watt hour (Wh) efficiency

$$\text{Wh efficiency (\%)} = \frac{\text{Wh output}}{\text{Wh input}} \times 100$$

Charging current

$$I = \frac{V - E}{R + r}$$

where:

V = supply voltage

E = e.m.f. of cell

R = resistance of series resistor and leads

r = internal resistance of cell

ILLUMINATION

Luminous flux

$$F = \frac{E \times a}{\mathrm{MF} \times \mathrm{CU}}$$

where:

E = illuminance in lux or lm/m^2
a = area in m^2
MF = maintenance factor
CU = coefficient of utilization

Illuminance

$$E = \frac{I}{d^2}$$

where:

I = luminous intensity in candelas
d = vertical distance from source in metres

also $E = I\dfrac{cos^3\theta}{d^2}$

where:

θ = some angle from the vertical in degrees